Textbook of Physics

Textbook of Physics

Summer Nolan

Larsen & Keller
www.larsen-keller.com

Textbook of Physics
Summer Nolan
ISBN: 978-1-64172-431-9 (Hardback)

 Larsen & Keller

Published by Larsen and Keller Education,
5 Penn Plaza,
19th Floor,
New York, NY 10001, USA

Cataloging-in-Publication Data

Textbook of physics / Summer Nolan.
    p. cm.
Includes bibliographical references and index.
ISBN 978-1-64172-431-9
1. Physics. 2. Physical sciences. 3. Dynamics. I. Nolan, Summer.
QC6 .T49 2020
530--dc23

For more information regarding Larsen and Keller Education and its products, please visit the publisher's website www.larsen-keller.com

Table of Contents

Permissions

Index

Preface

This book aims to help a broader range of students by exploring a wide variety of significant topics related to this discipline. It will help students in achieving a higher level of understanding of the subject and excel in their respective fields. This book would not have been possible without the unwavered support of my senior professors who took out the time to provide me feedback and help me with the process. I would also like to thank my family for their patience and support.

Physics is a natural science that deals with the study of matter. It primarily studies the motion and behavior of matter through space and time. It also investigates the related entities of energy and force. It is one of the fundamental fields of science that aims to understand the working and behavior of the universe. There are numerous sub-fields within this discipline. These are broadly categorized into classical physics and modern physics. Classical physics deals with matter and forces on a normal scale of observation. Modern physics studies the behavior of energy and matter under extreme conditions, or on very small or large scale. This book is a valuable compilation of topics, ranging from the basic to the most complex theories and principles in the field of physics. It attempts to understand the multiple branches that fall under the discipline of physics and how such concepts have practical applications. Those in search of information to further their knowledge will be greatly assisted by this book.

A brief overview of the book contents is provided below:

Chapter – Physics and its Branches

Physics is a natural science that deals with the study of matter. It focuses on the study of its motion and behaviour through space and time. It is classified into modern physics, nuclear physics, Atomic physics, optics, etc. These diverse branches of physics have been thoroughly discussed in this chapter.

Chapter – Force, Motion and Energy

Force is an interaction that changes the motion of the object. Motion is the action of changing the position of an object in a given interval of time. The quantitative property that is transferred to an object to perform work or to heat is referred as energy. The chapter closely examines key concepts of force, motion and energy to provide an extensive understanding of the subject.

Chapter – Light

Light is a form of electromagnetic radiation which requires no material medium to propagate. It illuminates any object on which it falls. The three main phenomena that form laws of light are reflection, refraction and diffraction of light. All these diverse principles of light have been carefully analysed in this chapter.

Chapter – Sound

Sound is a longitudinal wave that requires material medium such as solid, liquid or gas, to propagate. The reflection of direct sound that can be heard after a short interval of time is referred to as echo. This chapter has been carefully written to provide in-depth knowledge of the varied facets of sound.

Chapter – Fundamental Theories of Physics

Ample number of theories are used to study the concepts and applications related to physics. Fundamental theories of physics include Quantum field theory, M-theory, effective field theory, lattice field theory, etc. This chapter discusses these fundamental theories related to physics in detail.

Chapter – Laws of Physics

Laws of physics are derived from emperical observations. Laws such as Stefan-Boltzmann law, Pascal's law, Hooke's law, Charles's law and Boyle's law are among the basic laws of physics. The topics elaborated in this chapter will help in gaining a better perspective about these laws of physics.

Summer Nolan

1

Physics and its Branches

Physics is a natural science that deals with the study of matter. It focuses on the study of its motion and behaviour through space and time. It is classified into modern physics, nuclear physics, Atomic physics, optics, etc. These diverse branches of physics have been thoroughly discussed in this chapter.

PHYSICS

Physics is the branch of science that deals with the structure of matter and the interactions between the fundamental constituents of the observable universe. In the broadest sense, physics (from the Greek physikos) is concerned with all aspects of nature on both the macroscopic and submicroscopic levels. Its scope of study encompasses not only the behaviour of objects under the action of given forces but also the nature and origin of gravitational, electromagnetic, and nuclear force fields. Its ultimate objective is the formulation of a few comprehensive principles that bring together and explain all such disparate phenomena.

Physics is the basic physical science. Until rather recent times physics and natural philosophy were used interchangeably for the science whose aim is the discovery and formulation of the fundamental laws of nature. As the modern sciences developed and became increasingly specialized, physics came to denote that part of physical science not included in astronomy, chemistry, geology, and engineering. Physics plays an important role in all the natural sciences, however, and all such fields have branches in which physical laws and measurements receive special emphasis, bearing such names as astrophysics, geophysics, biophysics, and even psychophysics. Physics can, at base, be defined as the science of matter, motion, and energy. Its laws are typically expressed with economy and precision in the language of mathematics.

Both experiment, the observation of phenomena under conditions that are controlled as precisely as possible, and theory, the formulation of a unified conceptual framework, play essential and complementary roles in the advancement of physics. Physical experiments result in measurements, which are compared with the outcome predicted by theory. A theory that reliably predicts the results of experiments to which it is applicable is said to embody a law of physics. However, a law is always subject to modification, replacement, or restriction to a more limited domain, if a later experiment makes it necessary.

The ultimate aim of physics is to find a unified set of laws governing matter, motion, and energy at small (microscopic) subatomic distances, at the human (macroscopic) scale of everyday life, and out to the largest distances (e.g., those on the extragalactic scale). This ambitious goal has been realized to a notable extent. Although a completely unified theory of physical phenomena has not yet been achieved (and possibly never will be), a remarkably small set of fundamental physical laws appears able to account for all known phenomena. The body of physics developed up to about the turn of the 20th century, known as classical physics, can largely account for the motions of macroscopic objects that move slowly with respect to the speed of light and for such phenomena as heat, sound, electricity, magnetism, and light. The modern developments of relativity and quantum mechanics modify these laws insofar as they apply to higher speeds, very massive objects, and to the tiny elementary constituents of matter, such as electrons, protons, and neutrons.

The Scope Of Physics

The traditionally organized branches or fields of classical and modern physics are delineated below.

Mechanics

Mechanics is generally taken to mean the study of the motion of objects (or their lack of motion) under the action of given forces. Classical mechanics is sometimes considered a branch of applied mathematics. It consists of kinematics, the description of motion, and dynamics, the study of the action of forces in producing either motion or static equilibrium (the latter constituting the science of statics). The 20th-century subjects of quantum mechanics, crucial to treating the structure of matter, subatomic particles, superfluidity, superconductivity, neutron stars, and other major phenomena, and relativistic mechanics, important when speeds approach that of light, are forms of mechanics.

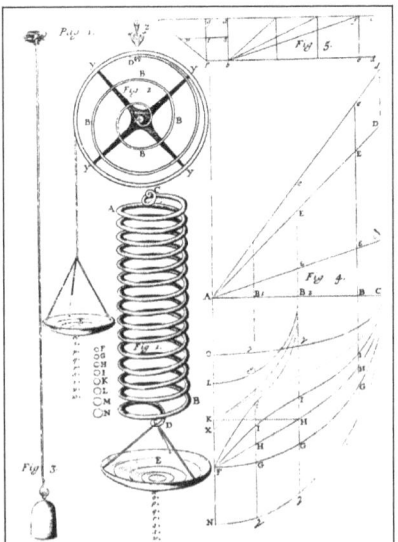

Illustration of Hooke's law of elasticity of materials, showing the stretching of a spring in proportion to the applied force, from Robert Hooke's Lectures de Potentia Restitutiva.

In classical mechanics the laws are initially formulated for point particles in which the dimensions, shapes, and other intrinsic properties of bodies are ignored. Thus in the first approximation even objects as large as the Earth and the Sun are treated as pointlike—e.g., in calculating planetary orbital motion. In rigid-body dynamics, the extension of bodies and their mass distributions are considered as well, but they are imagined to be incapable of deformation. The mechanics of deformable solids is elasticity; hydrostatics and hydrodynamics treat, respectively, fluids at rest and in motion.

The three laws of motion set forth by Isaac Newton form the foundation of classical mechanics, together with the recognition that forces are directed quantities (vectors) and combine accordingly. The first law, also called the law of inertia, states that, unless acted upon by an external force, an object at rest remains at rest, or if in motion, it continues to move in a straight line with constant speed. Uniform motion therefore does not require a cause. Accordingly, mechanics concentrates not on motion as such but on the change in the state of motion of an object that results from the net force acting upon it. Newton's second law equates the net force on an object to the rate of change of its momentum, the latter being the product of the mass of a body and its velocity. Newton's third law, that of action and reaction, states that when two particles interact, the forces each exerts on the other are equal in magnitude and opposite in direction. Taken together, these mechanical laws in principle permit the determination of the future motions of a set of particles, providing their state of motion is known at some instant, as well as the forces that act between them and upon them from the outside. From this deterministic character of the laws of classical mechanics, profound (and probably incorrect) philosophical conclusions have been drawn in the past and even applied to human history.

Lying at the most basic level of physics, the laws of mechanics are characterized by certain symmetry properties, as exemplified in the aforementioned symmetry between action and reaction forces. Other symmetries, such as the invariance (i.e., unchanging form) of the laws under reflections and rotations carried out in space, reversal of time, or transformation to a different part of space or to a different epoch of time, are present both in classical mechanics and in relativistic mechanics, and with certain restrictions, also in quantum mechanics. The symmetry properties of the theory can be shown to have as mathematical consequences basic principles known as conservation laws, which assert the constancy in time of the values of certain physical quantities under prescribed conditions. The conserved quantities are the most important ones in physics; included among them are mass and energy (in relativity theory, mass and energy are equivalent and are conserved together), momentum, angular momentum, and electric charge.

The Study of Gravitation

This field of inquiry has in the past been placed within classical mechanics for historical reasons, because both fields were brought to a high state of perfection by Newton and also because of its universal character. Newton's gravitational law states that every material particle in the universe attracts every other one with a force that acts along the line joining them and whose strength is directly proportional to the product of their masses and inversely proportional to the square of their separation. Newton's detailed accounting for the orbits of the planets and the Moon, as well as for such subtle gravitational effects as the tides and the precession of the equinoxes (a slow cyclical change in direction of the Earth's axis of rotation) through this fundamental force was the first triumph of classical mechanics. No further principles are required to understand the principal aspects of rocketry and space flight (although, of course, a formidable technology is needed to carry them out).

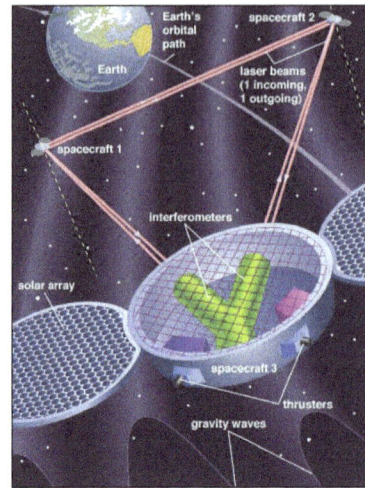

Laser Interferometer Space Antenna (LISA).

LISA, a Beyond Einstein Great Observatory, is scheduled for launch in 2015. Jointly funded by the National Aeronautics and Space Administration (NASA) and the European Space Agency (ESA), LISA will consist of three identical spacecraft that will trail the Earth in its orbit by about 50 million km (30 million miles). The spacecraft will contain thrusters for maneuvering them into an equilateral triangle, with sides of approximately 5 million km (3 million miles), such that the triangle's centre will be located along the Earth's orbit. By measuring the transmission of laser signals between the spacecraft (essentially a giant Michelson interferometer in space), scientists hope to detect and accurately measure gravity waves.

The modern theory of gravitation was formulated by Albert Einstein and is called the general theory of relativity. From the long-known equality of the quantity "mass" in Newton's second law of motion and that in his gravitational law, Einstein was struck by the fact that acceleration can locally annul a gravitational force (as occurs in the so-called weightlessness of astronauts in an Earth-orbiting spacecraft) and was led thereby to the concept of curved space-time. Completed in 1915, the theory was valued for many years mainly for its mathematical beauty and for correctly predicting a small number of phenomena, such as the gravitational bending of light around a massive object. Only in recent years, however, has it become a vital subject for both theoretical and experimental research. (Relativistic mechanics refers to Einstein's special theory of relativity, which is not a theory of gravitation).

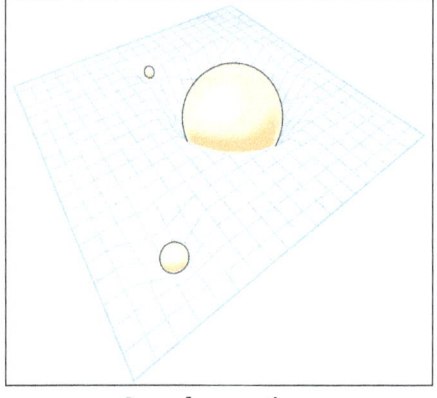

Curved space-time.

The four dimensional space-time continuum itself is distorted in the vicinity of any mass, with the amount of distortion depending on the mass and the distance from the mass. Thus, relativity accounts for Newton's inverse square law of gravity through geometry and thereby does away with the need for any mysterious "action at a distance."

The Study of Heat, Thermodynamics and Statistical Mechanics

Heat is a form of internal energy associated with the random motion of the molecular constituents of matter or with radiation. Temperature is an average of a part of the internal energy present in a body (it does not include the energy of molecular binding or of molecular rotation). The lowest possible energy state of a substance is defined as the absolute zero (−273.15 °C, or −459.67 °F) of temperature. An isolated body eventually reaches uniform temperature, a state known as thermal equilibrium, as do two or more bodies placed in contact. The formal study of states of matter at (or near) thermal equilibrium is called thermodynamics; it is capable of analyzing a large variety of thermal systems without considering their detailed microstructures.

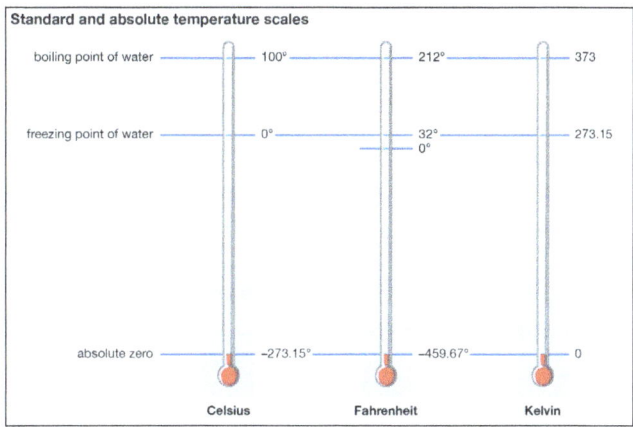

temperature scales.

First Law

The first law of thermodynamics is the energy conservation principle of mechanics (i.e., for all changes in an isolated system, the energy remains constant) generalized to include heat.

Second Law

The second law of thermodynamics asserts that heat will not flow from a place of lower temperature to one where it is higher without the intervention of an external device (e.g., a refrigerator). The concept of entropy involves the measurement of the state of disorder of the particles making up a system. For example, if tossing a coin many times results in a random-appearing sequence of heads and tails, the result has a higher entropy than if heads and tails tend to appear in clusters. Another formulation of the second law is that the entropy of an isolated system never decreases with time.

Third Law

The third law of thermodynamics states that the entropy at the absolute zero of temperature is zero, corresponding to the most ordered possible state.

Statistical Mechanics

The science of statistical mechanics derives bulk properties of systems from the mechanical properties of their molecular constituents, assuming molecular chaos and applying the laws of probability. Regarding each possible configuration of the particles as equally likely, the chaotic state (the state of maximum entropy) is so enormously more likely than ordered states that an isolated system will evolve to it, as stated in the second law of thermodynamics. Such reasoning, placed in mathematically precise form, is typical of statistical mechanics, which is capable of deriving the laws of thermodynamics but goes beyond them in describing fluctuations (i.e., temporary departures) from the thermodynamic laws that describe only average behaviour. An example of a fluctuation phenomenon is the random motion of small particles suspended in a fluid, known as Brownian motion.

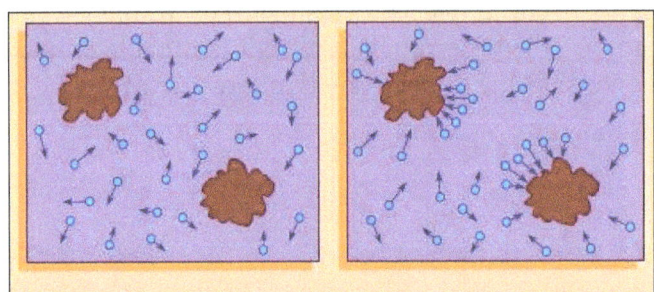

(Left) Random motion of a Brownian particle; (right) random discrepancy between the molecular pressures on different surfaces of the particle that cause motion.

Quantum statistical mechanics plays a major role in many other modern fields of science, as, for example, in plasma physics (the study of fully ionized gases), in solid-state physics, and in the study of stellar structure. From a microscopic point of view the laws of thermodynamics imply that, whereas the total quantity of energy of any isolated system is constant, what might be called the quality of this energy is degraded as the system moves inexorably, through the operation of the laws of chance, to states of increasing disorder until it finally reaches the state of maximum disorder (maximum entropy), in which all parts of the system are at the same temperature, and none of the state's energy may be usefully employed. When applied to the universe as a whole, considered as an isolated system, this ultimate chaotic condition has been called the "heat death."

The Study of Electricity and Magnetism

Although conceived of as distinct phenomena until the 19th century, electricity and magnetism are now known to be components of the unified field of electromagnetism. Particles with electric charge interact by an electric force, while charged particles in motion produce and respond to magnetic forces as well. Many subatomic particles, including the electrically charged electron and proton and the electrically neutral neutron, behave like elementary magnets. On the other hand, in spite of systematic searches undertaken, no magnetic monopoles, which would be the magnetic analogues of electric charges, have ever been found.

The field concept plays a central role in the classical formulation of electromagnetism, as well as in many other areas of classical and contemporary physics. Einstein's gravitational field, for example, replaces Newton's concept of gravitational action at a distance. The field describing the electric force between a pair of charged particles works in the following manner: each particle creates an electric field in the space surrounding it, and so also at the position occupied by the other particle; each particle responds to the force exerted upon it by the electric field at its own position.

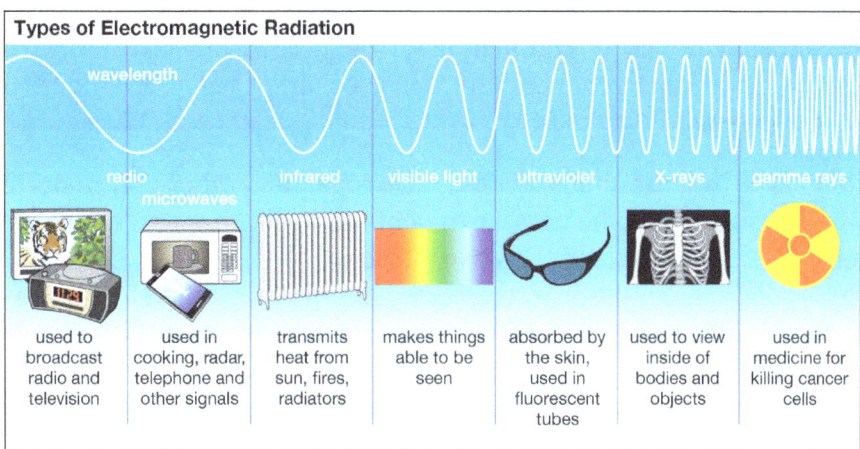

Radio waves, infrared rays, visible light, ultraviolet rays, X-rays, and gamma rays are all types of electromagnetic radiation. Radio waves have the longest wavelength, and gamma rays have the shortest wavelength.

Classical electromagnetism is summarized by the laws of action of electric and magnetic fields upon electric charges and upon magnets and by four remarkable equations formulated in the latter part of the 19th century by the Scottish physicist James Clerk Maxwell. The latter equations describe the manner in which electric charges and currents produce electric and magnetic fields, as well as the manner in which changing magnetic fields produce electric fields, and vice versa. From these relations Maxwell inferred the existence of electromagnetic waves—associated electric and magnetic fields in space, detached from the charges that created them, traveling at the speed of light, and endowed with such "mechanical" properties as energy, momentum, and angular momentum. The light to which the human eye is sensitive is but one small segment of an electromagnetic spectrum that extends from long-wavelength radio waves to short-wavelength gamma rays and includes X-rays, microwaves, and infrared (or heat) radiation.

Optics

Spectrum of white light by a diffraction grating. With a prism, the red end of the spectrum is more compressed than the violet end.

Because light consists of electromagnetic waves, the propagation of light can be regarded as merely a branch of electromagnetism. However, it is usually dealt with as a separate subject called optics: the part that deals with the tracing of light rays is known as geometrical optics, while the part that treats the distinctive wave phenomena of light is called physical optics. More recently, there has developed a new and vital branch, quantum optics, which is concerned with the theory and application of the laser, a device that produces an intense coherent beam of unidirectional radiation useful for many applications.

The formation of images by lenses, microscopes, telescopes, and other optical devices is described by ray optics, which assumes that the passage of light can be represented by straight lines, that is, rays. The subtler effects attributable to the wave property of visible light, however, require the explanations of physical optics. One basic wave effect is interference, whereby two waves present in a region of space combine at certain points to yield an enhanced resultant effect (e.g., the crests of the component waves adding together); at the other extreme, the two waves can annul each other, the crests of one wave filling in the troughs of the other. Another wave effect is diffraction, which causes light to spread into regions of the geometric shadow and causes the image produced by any optical device to be fuzzy to a degree dependent on the wavelength of the light. Optical instruments such as the interferometer and the diffraction grating can be used for measuring the wavelength of light precisely (about 500 micrometres) and for measuring distances to a small fraction of that length.

Atomic and Chemical Physics

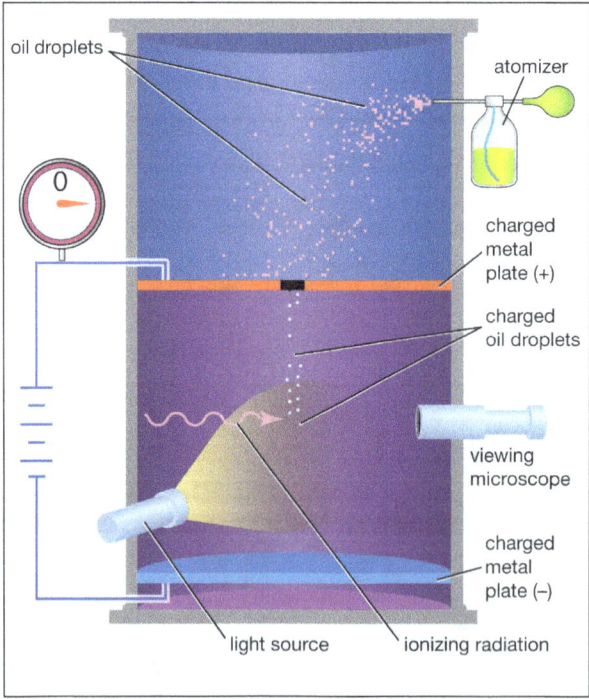

Millikan oil-drop experiment.

One of the great achievements of the 20th century was the establishment of the validity of the atomic hypothesis, first proposed in ancient times, that matter is made up of relatively few kinds of small, identical parts—namely, atoms. However, unlike the indivisible atom of Democritus and other ancients, the atom, as it is conceived today, can be separated into constituent electrons and nucleus. Atoms combine to form molecules, whose structure is studied by chemistry and physical chemistry; they also form other types of compounds, such as crystals, studied in the field of condensed-matter physics. Such disciplines study the most important attributes of matter (not excluding biologic matter) that are encountered in normal experience—namely, those that depend almost entirely on the outer parts of the electronic structure of atoms. Only the mass of the atomic nucleus and its charge, which is equal to the total charge of the electrons in the neutral atom, affect the chemical and physical properties of matter.

Between 1909 and 1910 the American physicist Robert Millikan conducted a series of oil-drop experiments. By comparing applied electric force with changes in the motion of the oil drops, he was able to determine the electric charge on each drop. He found that all of the drops had charges that were simple multiples of a single number, the fundamental charge of the electron.

Although there are some analogies between the solar system and the atom due to the fact that the strengths of gravitational and electrostatic forces both fall off as the inverse square of the distance, the classical forms of electromagnetism and mechanics fail when applied to tiny, rapidly moving atomic constituents. Atomic structure is comprehensible only on the basis of quantum mechanics, and its finer details require as well the use of quantum electrodynamics (QED).

Atomic properties are inferred mostly by the use of indirect experiments. Of greatest importance has been spectroscopy, which is concerned with the measurement and interpretation of the electromagnetic radiations either emitted or absorbed by materials. These radiations have a distinctive character, which quantum mechanics relates quantitatively to the structures that produce and absorb them. It is truly remarkable that these structures are in principle, and often in practice, amenable to precise calculation in terms of a few basic physical constants: the mass and charge of the electron, the speed of light, and Planck's constant (approximately $6.62606957 \times 10{-34}$ joule·second), the fundamental constant of the quantum theory named for the German physicist Max Planck.

Condensed-matter Physics

This field, which treats the thermal, elastic, electrical, magnetic, and optical properties of solid and liquid substances, grew at an explosive rate in the second half of the 20th century and scored numerous important scientific and technical achievements, including the transistor. Among solid materials, the greatest theoretical advances have been in the study of crystalline materials whose simple repetitive geometric arrays of atoms are multiple-particle systems that allow treatment by quantum mechanics. Because the atoms in a solid are coordinated with each other over large distances, the theory must go beyond that appropriate for atoms and molecules. Thus conductors, such as metals, contain some so-called free electrons, or valence electrons, which are responsible for the electrical and most of the thermal conductivity of the material and which belong collectively to the whole solid rather than to individual atoms. Semiconductors and insulators, either crystalline or amorphous, are other materials studied in this field of physics.

Other aspects of condensed matter involve the properties of the ordinary liquid state, of liquid crystals, and, at temperatures near absolute zero, of the so-called quantum liquids. The latter exhibit a property known as superfluidity (completely frictionless flow), which is an example of macroscopic quantum phenomena. Such phenomena are also exemplified by superconductivity (completely resistance-less flow of electricity), a low-temperature property of certain metallic and ceramic materials. Besides their significance to technology, macroscopic liquid and solid quantum states are important in astrophysical theories of stellar structure in, for example, neutron stars.

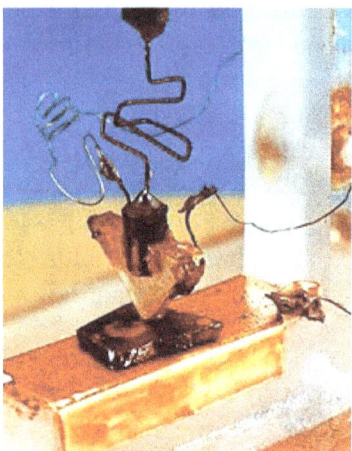

The first transistor, invented by American physicists John Bardeen,
Walter H. Brattain, and William B. Shockley.

Nuclear Physics

This branch of physics deals with the structure of the atomic nucleus and the radiation from unstable nuclei. About 10,000 times smaller than the atom, the constituent particles of the nucleus, protons and neutrons, attract one another so strongly by the nuclear forces that nuclear energies are approximately 1,000,000 times larger than typical atomic energies. Quantum theory is needed for understanding nuclear structure.

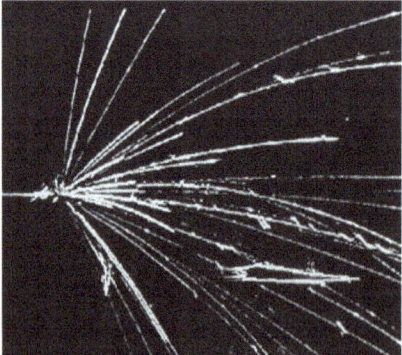

Particle tracks from the collision of an accelerated nucleus of a niobium atom with another niobium nucleus.
The single line on the left is the track of the incoming projectile nucleus, and the other
tracks are fragments from the collision.

Like excited atoms, unstable radioactive nuclei (either naturally occurring or artificially produced) can emit electromagnetic radiation. The energetic nuclear photons are called gamma rays. Radioactive nuclei also emit other particles: negative and positive electrons (beta rays), accompanied by neutrinos, and helium nuclei (alpha rays).

A principal research tool of nuclear physics involves the use of beams of particles (e.g., protons or electrons) directed as projectiles against nuclear targets. Recoiling particles and any resultant nuclear fragments are detected, and their directions and energies are analyzed to reveal details of nuclear structure and to learn more about the strong force. A much weaker nuclear force, the so-called weak interaction, is responsible for the emission of beta rays. Nuclear collision experiments use beams of higher-energy particles, including those of unstable particles called mesons produced by primary nuclear collisions in accelerators dubbed meson factories. Exchange of mesons between protons and neutrons is directly responsible for the strong force.

In radioactivity and in collisions leading to nuclear breakup, the chemical identity of the nuclear target is altered whenever there is a change in the nuclear charge. In fission and fusion nuclear reactions in which unstable nuclei are, respectively, split into smaller nuclei or amalgamated into larger ones, the energy release far exceeds that of any chemical reaction.

Particle Physics

One of the most significant branches of contemporary physics is the study of the fundamental subatomic constituents of matter, the elementary particles. This field, also called high-energy physics, emerged in the 1930s out of the developing experimental areas of nuclear and cosmic-ray physics. Initially investigators studied cosmic rays, the very-high-energy extraterrestrial radiations that fall upon the Earth and interact in the atmosphere. However, after World War II, scientists gradually began using high-energy particle accelerators to provide subatomic particles for study. Quantum field theory, a generalization of QED to other types of force fields, is essential for the analysis of high-energy physics. Subatomic particles cannot be visualized as tiny analogues of ordinary material objects such as billiard balls, for they have properties that appear contradictory from the classical viewpoint. That is to say, while they possess charge, spin, mass, magnetism, and other complex characteristics, they are nonetheless regarded as pointlike.

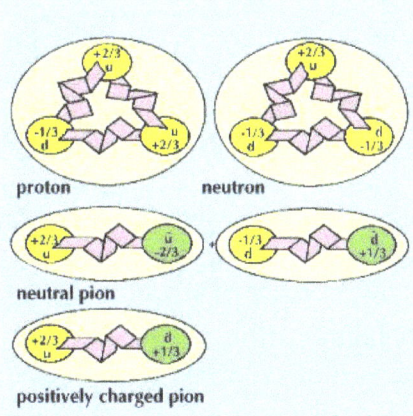

Very simplified illustrations of protons, neutrons, pions, and other hadrons show that they are made of quarks (yellow spheres) and antiquarks (green spheres), which are bound together by gluons (bent ribbons).

During the latter half of the 20th century, a coherent picture evolved of the underlying strata of matter involving two types of subatomic particles: fermions (baryons and leptons), which have odd half-integral angular momentum (spin 1/2, 3/2) and make up ordinary matter; and bosons (gluons, mesons, and photons), which have integral spins and mediate the fundamental forces of physics. Leptons (e.g., electrons, muons, taus), gluons, and photons are believed to be truly fundamental particles. Baryons (e.g., neutrons, protons) and mesons (e.g., pions, kaons), collectively known as hadrons, are believed to be formed from indivisible elements known as quarks, which have never been isolated.

Quarks come in six types, or "flavours," and have matching antiparticles, known as antiquarks. Quarks have charges that are either positive two-thirds or negative one-third of the electron's charge, while antiquarks have the opposite charges. Like quarks, each lepton has an antiparticle with properties that mirror those of its partner (the antiparticle of the negatively charged electron is the positive electron, or positron; that of the neutrino is the antineutrino). In addition to their electric and magnetic properties, quarks participate in both the strong force (which binds them together) and the weak force (which underlies certain forms of radioactivity), while leptons take part in only the weak force.

Baryons, such as neutrons and protons, are formed by combining three quarks—thus baryons have a charge of −1, 0, or 1. Mesons, which are the particles that mediate the strong force inside the atomic nucleus, are composed of one quark and one antiquark; all known mesons have a charge of −2, −1, 0, 1, or 2. Most of the possible quark combinations, or hadrons, have very short lifetimes, and many of them have never been seen, though additional ones have been observed with each new generation of more powerful particle accelerators.

The quantum fields through which quarks and leptons interact with each other and with themselves consist of particle-like objects called quanta (from which quantum mechanics derives its name). The first known quanta were those of the electromagnetic field; they are also called photons because light consists of them. A modern unified theory of weak and electromagnetic interactions, known as the electroweak theory, proposes that the weak force involves the exchange of particles about 100 times as massive as protons. These massive quanta have been observed—namely, two charged particles, W+ and W−, and a neutral one, W0.

In the theory of the strong force known as quantum chromodynamics (QCD), eight quanta, called gluons, bind quarks to form baryons and also bind quarks to antiquarks to form mesons, the force itself being dubbed the "colour force." (This unusual use of the term colour is a somewhat forced analogue of ordinary colour mixing.). Quarks are said to come in three colours—red, blue, and green. (The opposites of these imaginary colours, minus-red, minus-blue, and minus-green, are ascribed to antiquarks.) Only certain colour combinations, namely colour-neutral, or "white" (i.e., equal mixtures of the above colours cancel out one another, resulting in no net colour), are conjectured to exist in nature in an observable form. The gluons and quarks themselves, being coloured, are permanently confined (deeply bound within the particles of which they are a part), while the colour-neutral composites such as protons can be directly observed. One consequence of colour confinement is that the observable particles are either electrically neutral or have charges that are integral multiples of the charge of the electron. A number of specific predictions of QCD have been experimentally tested and found correct.

Quantum Mechanics

Although the various branches of physics differ in their experimental methods and theoretical approaches, certain general principles apply to all of them. The forefront of contemporary advances in physics lies in the submicroscopic regime, whether it be in atomic, nuclear, condensed-matter, plasma, or particle physics, or in quantum optics, or even in the study of stellar structure. All are based upon quantum theory (i.e., quantum mechanics and quantum field theory) and relativity, which together form the theoretical foundations of modern physics. Many physical quantities whose classical counterparts vary continuously over a range of possible values are in quantum theory constrained to have discontinuous, or discrete, values. Furthermore, the intrinsically deterministic character of values in classical physics is replaced in quantum theory by intrinsic uncertainty.

According to quantum theory, electromagnetic radiation does not always consist of continuous waves; instead it must be viewed under some circumstances as a collection of particle-like photons, the energy and momentum of each being directly proportional to its frequency (or inversely proportional to its wavelength, the photons still possessing some wavelike characteristics). Conversely, electrons and other objects that appear as particles in classical physics are endowed by quantum theory with wavelike properties as well, such a particle's quantum wavelength being inversely proportional to its momentum. In both instances, the proportionality constant is the characteristic quantum of action (action being defined as energy × time)—that is to say, Planck's constant divided by 2π, or $\hbar$.

In principle, all of atomic and molecular physics, including the structure of atoms and their dynamics, the periodic table of elements and their chemical behaviour, as well as the spectroscopic, electrical, and other physical properties of atoms, molecules, and condensed matter, can be accounted for by quantum mechanics. Roughly speaking, the electrons in the atom must fit around the nucleus as some sort of standing wave (as given by the Schrödinger equation) analogous to the waves on a plucked violin or guitar string. As the fit determines the wavelength of the quantum wave, it necessarily determines its energy state. Consequently, atomic systems are restricted to certain discrete, or quantized, energies. When an atom undergoes a discontinuous transition, or quantum jump, its energy changes abruptly by a sharply defined amount, and a photon of that energy is emitted when the energy of the atom decreases, or is absorbed in the opposite case.

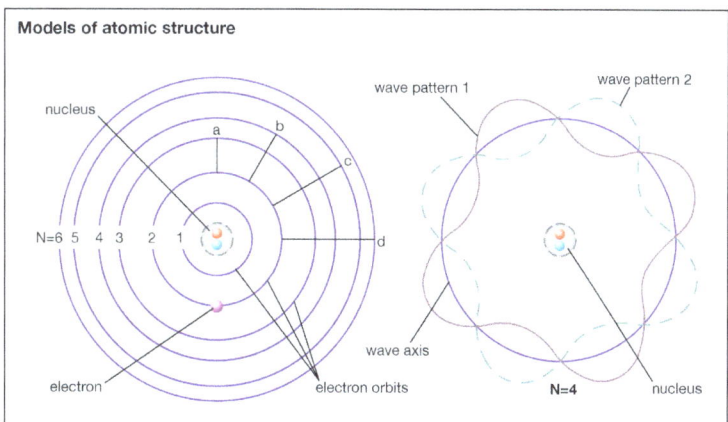

The Bohr theory sees an electron (left) as a point mass occupying certain energy levels. Wave mechanics sees an electron as a wave washing back and forth in the atom in certain patterns only. The wave patterns and energy levels correspond exactly.

Although atomic energies can be sharply defined, the positions of the electrons within the atom cannot be, quantum mechanics giving only the probability for the electrons to have certain locations. This is a consequence of the feature that distinguishes quantum theory from all other approaches to physics, the uncertainty principle of the German physicist Werner Heisenberg. This principle holds that measuring a particle's position with increasing precision necessarily increases the uncertainty as to the particle's momentum, and conversely. The ultimate degree of uncertainty is controlled by the magnitude of Planck's constant, which is so small as to have no apparent effects except in the world of microstructures. In the latter case, however, because both a particle's position and its velocity or momentum must be known precisely at some instant in order to predict its future history, quantum theory precludes such certain prediction and thus escapes determinism.

The complementary wave and particle aspects, or wave–particle duality, of electromagnetic radiation and of material particles furnish another illustration of the uncertainty principle. When an electron exhibits wavelike behaviour, as in the phenomenon of electron diffraction, this excludes its exhibiting particle-like behaviour in the same observation. Similarly, when electromagnetic radiation in the form of photons interacts with matter, as in the Compton effect in which X-ray photons collide with electrons, the result resembles a particle-like collision and the wave nature of electromagnetic radiation is precluded. The principle of complementarity, asserted by the Danish physicist Niels Bohr, who pioneered the theory of atomic structure, states that the physical world presents itself in the form of various complementary pictures, no one of which is by itself complete, all of these pictures being essential for our total understanding. Thus both wave and particle pictures are needed for understanding either the electron or the photon.

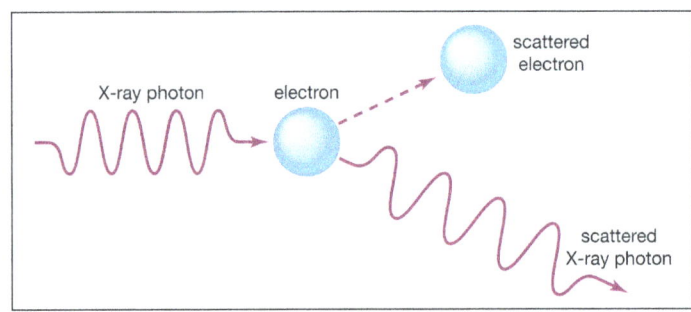

The Compton effect.

When a beam of X-rays is aimed at a target material, some of the beam is deflected, and the scattered X-rays have a greater wavelength than the original beam. The physicist Arthur Holly Compton concluded that this phenomenon could only be explained if the X-rays were understood to be made up of discrete bundles or particles, now called photons, that lost some of their energy in the collisions with electrons in the target material and then scattered at lower energy.

Although it deals with probabilities and uncertainties, the quantum theory has been spectacularly successful in explaining otherwise inaccessible atomic phenomena and in thus far meeting every experimental test. Its predictions, especially those of QED, are the most precise and the best checked of any in physics; some of them have been tested and found accurate to better than one part per billion.

Relativistic Mechanics

In classical physics, space is conceived as having the absolute character of an empty stage in which events in nature unfold as time flows onward independently; events occurring simultaneously for one observer are presumed to be simultaneous for any other; mass is taken as impossible to create or destroy; and a particle given sufficient energy acquires a velocity that can increase without limit. The special theory of relativity, developed principally by Albert Einstein in 1905 and now so adequately confirmed by experiment as to have the status of physical law, shows that all these, as well as other apparently obvious assumptions, are false.

Specific and unusual relativistic effects flow directly from Einstein's two basic postulates, which are formulated in terms of so-called inertial reference frames. These are reference systems that move in such a way that in them Isaac Newton's first law, the law of inertia, is valid. The set of inertial frames consists of all those that move with constant velocity with respect to each other (accelerating frames therefore being excluded). Einstein's postulates are: (1) All observers, whatever their state of motion relative to a light source, measure the same speed for light; and (2) The laws of physics are the same in all inertial frames.

The first postulate, the constancy of the speed of light, is an experimental fact from which follow the distinctive relativistic phenomena of space contraction (or Lorentz-FitzGerald contraction), time dilation, and the relativity of simultaneity: as measured by an observer assumed to be at rest, an object in motion is contracted along the direction of its motion, and moving clocks run slow; two spatially separated events that are simultaneous for a stationary observer occur sequentially for a moving observer. As a consequence, space intervals in three-dimensional space are related to time intervals, thus forming so-called four-dimensional space-time.

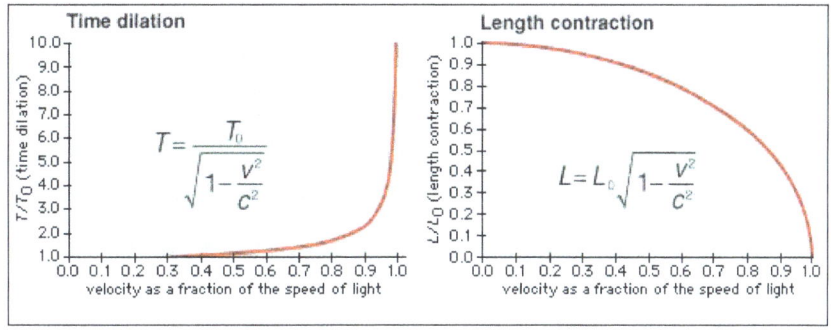

Length contraction and time dilation.

As an object approaches the speed of light, an observer sees the object become shorter and its time interval become longer, relative to the length and time interval when the object is at rest.

The second postulate is called the principle of relativity. It is equally valid in classical mechanics (but not in classical electrodynamics until Einstein reinterpreted it). This postulate implies, for example, that table tennis played on a train moving with constant velocity is just like table tennis played with the train at rest, the states of rest and motion being physically indistinguishable. In relativity theory, mechanical quantities such as momentum and energy have forms that are different from their classical counterparts but give the same values for speeds that are small compared to the speed of light, the maximum permissible speed in nature (about 300,000 kilometres per second, or 186,000 miles per second). According to relativity, mass and energy are equivalent and interchangeable quantities, the equivalence being expressed by Einstein's famous mass-energy equation E = mc2, where m is an object's mass and c is the speed of light.

The general theory of relativity is Einstein's theory of gravitation, which uses the principle of the equivalence of gravitation and locally accelerating frames of reference. Einstein's theory has special mathematical beauty; it generalizes the "flat" space-time concept of special relativity to one of curvature. It forms the background of all modern cosmological theories. In contrast to some vulgarized popular notions of it, which confuse it with moral and other forms of relativism, Einstein's theory does not argue that "all is relative." On the contrary, it is largely a theory based upon those physical attributes that do not change, or, in the language of the theory, that are invariant.

Conservation Laws and Symmetry

Since the early period of modern physics, there have been conservation laws, which state that certain physical quantities, such as the total electric charge of an isolated system of bodies, do not change in the course of time. In the 20th century it has been proved mathematically that such laws follow from the symmetry properties of nature, as expressed in the laws of physics. The conservation of mass-energy of an isolated system, for example, follows from the assumption that the laws of physics may depend upon time intervals but not upon the specific time at which the laws are applied. The symmetries and the conservation laws that follow from them are regarded by modern physicists as being even more fundamental than the laws themselves, since they are able to limit the possible forms of laws that may be proposed in the future.

Conservation laws are valid in classical, relativistic, and quantum theory for mass-energy, momentum, angular momentum, and electric charge. (In nonrelativistic physics, mass and energy are separately conserved.). Momentum, a directed quantity equal to the mass of a body multiplied by its velocity or to the total mass of two or more bodies multiplied by the velocity of their centre of mass, is conserved when, and only when, no external force acts. Similarly angular momentum, which is related to spinning motions, is conserved in a system upon which no net turning force, called torque, acts. External forces and torques break the symmetry conditions from which the respective conservation laws follow.

In quantum theory, and especially in the theory of elementary particles, there are additional symmetries and conservation laws, some exact and others only approximately valid, which play no significant role in classical physics. Among these are the conservation of so-called quantum numbers related to left-right reflection symmetry of space (called parity) and to the reversal symmetry of motion (called time reversal). These quantum numbers are conserved in all processes other than the weak force.

Other symmetry properties not obviously related to space and time (and referred to as internal symmetries) characterize the different families of elementary particles and, by extension, their composites. Quarks, for example, have a property called baryon number, as do protons, neutrons, nuclei, and unstable quark composites. All of these except the quarks are known as baryons. A failure of baryon-number conservation would exhibit itself, for instance, by a proton decaying into lighter non-baryonic particles. Indeed, intensive search for such proton decay has been conducted, but so far it has been fruitless. Similar symmetries and conservation laws hold for an analogously defined lepton number, and they also appear, as does the law of baryon conservation, to hold absolutely.

Fundamental Forces and Fields

The four basic forces of nature, in order of increasing strength, are thought to be: (1) the gravitational force between particles with mass; (2) the electromagnetic force between particles with charge or magnetism or both; (3) the colour force, or strong force, between quarks; and (4) the weak force by which, for example, quarks can change their type, so that a neutron decays into a proton, an electron, and an antineutrino. The strong force that binds protons and neutrons into nuclei and is responsible for fission, fusion, and other nuclear reactions is in principle derived from the colour force. Nuclear physics is thus related to QCD as chemistry is to atomic physics.

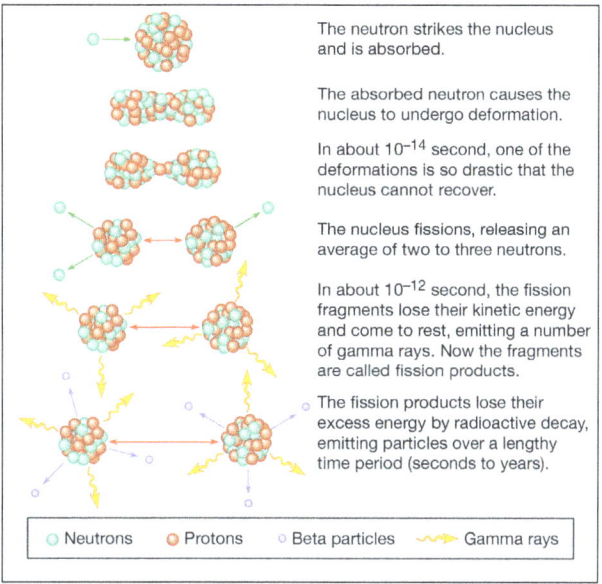

Sequence of events in the fission of a uranium nucleus by a neutron.

According to quantum field theory, each of the four fundamental interactions is mediated by the exchange of quanta, called vector gauge bosons, which share certain common characteristics. All have an intrinsic spin of one unit, measured in terms of Planck's constant $\hbar$. (Leptons and quarks each have one-half unit of spin.). Gauge theory studies the group of transformations, or Lie group, that leaves the basic physics of a quantum field invariant. Lie groups, which are named for the 19th-century Norwegian mathematician Sophus Lie, possess a special type of symmetry and continuity that made them first useful in the study of differential equations on smooth manifolds (an abstract mathematical space for modeling physical processes). This symmetry was first seen in the equations for electromagnetic potentials, quantities from which electromagnetic fields can be derived. It is possessed in pure form by the eight massless gluons of QCD, but in the electroweak theory—the unified theory of electromagnetic and weak force interactions—gauge symmetry is partially broken, so that only the photon remains massless, with the other gauge bosons (W^+, W^-, and Z) acquiring large masses. Theoretical physicists continue to seek a further unification of QCD with the electroweak theory and, more ambitiously still, to unify them with a quantum version of gravity in which the force would be transmitted by massless quanta of two units of spin called gravitons.

The Methodology of Physics

Physics has evolved and continues to evolve without any single strategy. Essentially an experimental science, refined measurements can reveal unexpected behaviour. On the other hand, mathematical extrapolation of existing theories into new theoretical areas, critical reexamination of apparently obvious but untested assumptions, argument by symmetry or analogy, aesthetic judgment, pure accident, and hunch—each of these plays a role (as in all of science). Thus, for example, the quantum hypothesis proposed by the German physicist Max Planck was based on observed departures of the character of blackbody radiation (radiation emitted by a heated body that absorbs all radiant energy incident upon it) from that predicted by classical electromagnetism. The English physicist P.A.M. Dirac predicted the existence of the positron in making a relativistic extension of the quantum theory of the electron. The elusive neutrino, without mass or charge, was hypothesized by the German physicist Wolfgang Pauli as an alternative to abandoning the conservation laws in the beta-decay process. Maxwell conjectured that if changing magnetic fields create electric fields (which was known to be so), then changing electric fields might create magnetic fields, leading him to the electromagnetic theory of light. Albert Einstein's special theory of relativity was based on a critical reexamination of the meaning of simultaneity, while his general theory of relativity rests on the equivalence of inertial and gravitational mass.

Although the tactics may vary from problem to problem, the physicist invariably tries to make unsolved problems more tractable by constructing a series of idealized models, with each successive model being a more realistic representation of the actual physical situation. Thus, in the theory of gases, the molecules are at first imagined to be particles that are as structureless as billiard balls with vanishingly small dimensions. This ideal picture is then improved on step by step.

The correspondence principle, a useful guiding principle for extending theoretical interpretations, was formulated by the Danish physicist Niels Bohr in the context of the quantum theory. It asserts that when a valid theory is generalized to a broader arena, the new theory's predictions must agree with the old one in the overlapping region in which both are applicable. For example, the more comprehensive theory of physical optics must yield the same result as the more restrictive theory of ray optics whenever wave effects proportional to the wavelength of light are negligible on account of the smallness of that wavelength. Similarly, quantum mechanics must yield the same results as classical mechanics in circumstances when Planck's constant can be considered as negligibly small. Likewise, for speeds small compared to the speed of light (as for baseballs in play), relativistic mechanics must coincide with Newtonian classical mechanics.

Some ways in which experimental and theoretical physicists attack their problems are illustrated by the following examples:

The modern experimental study of elementary particles began with the detection of new types of unstable particles produced in the atmosphere by primary radiation, the latter consisting mainly of high-energy protons arriving from space. The new particles were detected in Geiger counters and identified by the tracks they left in instruments called cloud chambers and in photographic plates. After World War II, particle physics, then known as high-energy nuclear physics, became a major field of science. Today's high-energy particle accelerators can be several kilometres in length, cost hundreds (or even thousands) of millions of dollars, and accelerate particles to enormous energies (trillions of electron volts). Experimental teams, such as those that discovered the W+, W−, and Z quanta of the weak force at the European Laboratory for Particle Physics (CERN) in Geneva, which is funded by its 20 European member states, can have 100 or more physicists from many countries, along with a larger number of technical workers serving as support personnel. A variety of visual and electronic techniques are used to interpret and sort the huge amounts of data produced by their efforts, and particle-physics laboratories are major users of the most advanced technology, be it superconductive magnets or supercomputers.

Theoretical physicists use mathematics both as a logical tool for the development of theory and for calculating predictions of the theory to be compared with experiment. Newton, for one, invented integral calculus to solve the following problem, which was essential to his formulation of the law of universal gravitation: Assuming that the attractive force between any pair of point particles is inversely proportional to the square of the distance separating them, how does a spherical distribution of particles, such as the Earth, attract another nearby object? Integral calculus, a procedure for summing many small contributions, yields the simple solution that the Earth itself acts as a point particle with all its mass concentrated at the centre. In modern physics, Dirac predicted the existence of the then-unknown positive electron (or positron) by finding an equation for the electron that would combine quantum mechanics and the special theory of relativity.

Relations between Physics and other Disciplines and Society

Influence of Physics on Related Disciplines

Because physics elucidates the simplest fundamental questions in nature on which there can be a consensus, it is hardly surprising that it has had a profound impact on other fields of science, on philosophy, on the worldview of the developed world, and, of course, on technology.

Indeed, whenever a branch of physics has reached such a degree of maturity that its basic elements are comprehended in general principles, it has moved from basic to applied physics and thence to technology. Thus almost all current activity in classical physics consists of applied physics, and its contents form the core of many branches of engineering. Discoveries in modern physics are converted with increasing rapidity into technical innovations and analytical tools for associated disciplines. There are, for example, such nascent fields as nuclear and biomedical engineering, quantum chemistry and quantum optics, and radio, X-ray, and gamma-ray astronomy, as well as such analytic tools as radioisotopes, spectroscopy, and lasers, which all stem directly from basic physics.

Apart from its specific applications, physics—especially Newtonian mechanics—has become the prototype of the scientific method. Its experimental and analytic methods sometimes being imitated (and sometimes inappropriately so) in fields far from the related physical sciences. Some of the organizational aspects of physics, based partly on the successes of the radar and atomic-bomb projects of World War II, also have been imitated in large-scale scientific projects, as, for example, in astronomy and space research.

The great influence of physics on the branches of philosophy concerned with the conceptual basis of human perceptions and understanding of nature, such as epistemology, is evidenced by the earlier designation of physics itself as natural philosophy. Present-day philosophy of science deals largely, though not exclusively, with the foundations of physics. Determinism, the philosophical doctrine that the universe is a vast machine operating with strict causality whose future is determined in all detail by its present state, is rooted in Newtonian mechanics, which obeys that principle. Moreover, the schools of materialism, naturalism, and empiricism have in large degree considered physics to be a model for philosophical inquiry. An extreme position is taken by the logical positivists, whose radical distrust of the reality of anything not directly observable leads them to demand that all significant statements must be formulated in the language of physics.

The uncertainty principle of quantum theory has prompted a reexamination of the question of determinism, and its other philosophical implications remain in doubt. Particularly problematic is the matter of the meaning of measurement, for which recent theories and experiments confirm some apparently noncausal predictions of standard quantum theory. It is fair to say that though physicists agree that quantum theory works, they still differ as to what it means.

Influence of Related Disciplines on Physics

The relationship of physics to its bordering disciplines is a reciprocal one. Just as technology feeds on fundamental science for new practical innovations, so physics appropriates the techniques and instrumentation of modern technology for advancing itself. Thus experimental physicists utilize increasingly refined and precise electronic devices. Moreover, they work closely with engineers in designing basic scientific equipment, such as high-energy particle accelerators. Mathematics has always been the primary tool of the theoretical physicist, and even abstruse fields of mathematics such as group theory and differential geometry have become invaluable to the theoretician classifying subatomic particles or investigating the symmetry characteristics of atoms and molecules. Much of contemporary research in physics depends on the high-speed computer. It allows the theoretician to perform computations that are too lengthy or complicated to be done with paper and pencil. Also, it allows experimentalists to incorporate the computer into their apparatus, so that the results of measurements can be provided nearly instantaneously on-line as summarized data while an experiment is in progress.

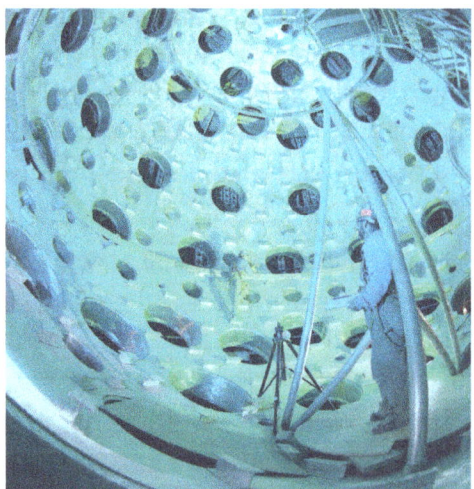

Laser-activated fusion.

Interior of the U.S. Department of Energy's National Ignition Facility (NIF), located at Lawrence Livermore National Laboratory, Livermore, California. The NIF target chamber uses a high-energy laser to heat fusion fuel to temperatures sufficient for thermonuclear ignition. The facility is used for basic science, fusion energy research, and nuclear weapons testing.

The Physicist in Society

Because of the remoteness of much of contemporary physics from ordinary experience and its reliance on advanced mathematics, physicists have sometimes seemed to the public to be initiates in a latter-day secular priesthood who speak an arcane language and can communicate their findings to laymen only with great difficulty. Yet, the physicist has come to play an increasingly significant role in society, particularly since World War II. Governments have supplied substantial funds for research at academic institutions and at government laboratories through such agencies as the National Science Foundation and the Department of Energy in the United States, which has also established a number of national laboratories, including the Fermi National Accelerator Laboratory in Batavia, Ill., with one of the world's largest particle accelerators. CERN is composed of 14 European countries and operates a large accelerator at the Swiss–French border. Physics research is supported in Germany by the Max Planck Society for the Advancement of Science and in Japan by the Japan Society for the Promotion of Science. In Trieste, Italy, there is the International Center for Theoretical Physics, which has strong ties to developing countries. These are only a few examples of the widespread international interest in fundamental physics.

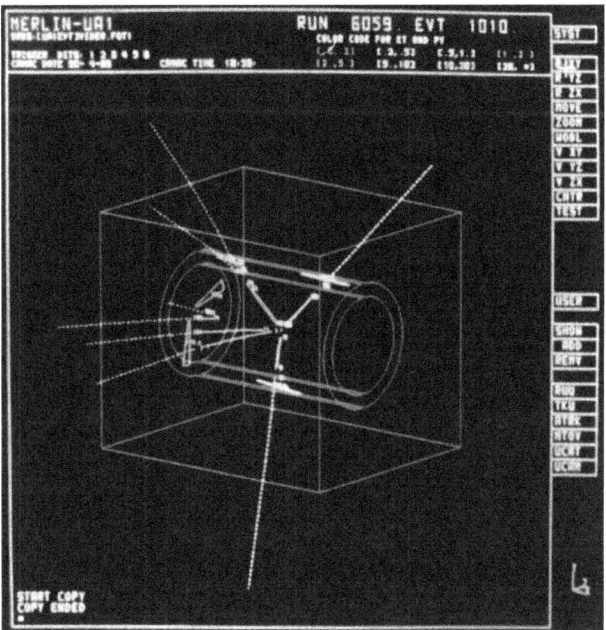

Tracks emerging from a proton-antiproton collision at the centre of the UA1 detector at CERN include those of an energetic electron (straight down) and a positron (upper right). These two particles have come from the decay of a Z0; when their energies are added together, the total is equal to the Z0's mass.

Basic research in physics is obviously dependent on public support and funding, and with this development has come, albeit slowly, a growing recognition within the physics community of the social responsibility of scientists for the consequences of their work and for the more general problems of science and society.

MODERN PHYSICS

Modern physics is an effort to understand the underlying processes of the interactions of matter utilizing the tools of science & engineering. In general, the term is used to refer to any branch of physics either developed in the early 20th century and onwards, or branches greatly influenced by early 20th century physics.

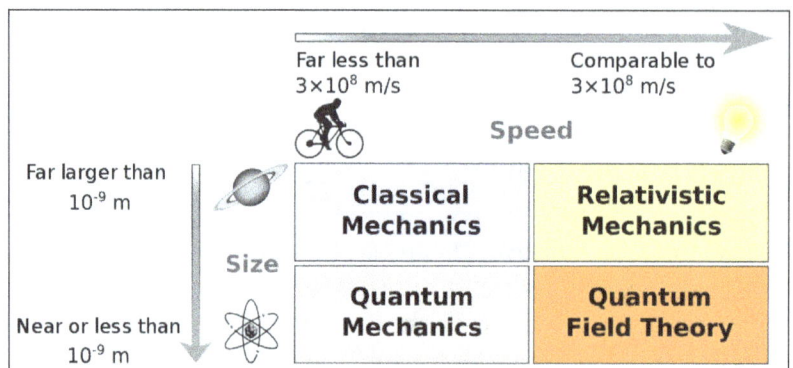

Classical physics is usually concerned with everyday conditions: speeds much lower than the speed of light, and sizes much greater than that of atoms. Modern physics is usually concerned with high velocities and small distances.

Small velocities and large distances is usually the realm of classical physics. Modern physics, however, often involves extreme conditions: Quantum effects typically involve distances comparable to atoms (roughly 10^{-9} m), while relativistic effects typically involve velocities comparable to the speed of light (roughly 3×10^8 m/s). In general, quantum and relativistic effects exist across all scales, although these effects can be very small in everyday life.

In a literal sense, the term *modern physics*, means up-to-date physics. In this sense, a significant portion of so-called *classical physics* is modern. However, since roughly 1890, new discoveries have caused significant paradigm shifts: the advent of quantum mechanics (QM) and of Einsteinian relativity (ER). Physics that incorporates elements of either QM or ER (or both) is said to be *modern physics*. It is in this latter sense that the term is generally used.

Modern physics is often encountered when dealing with extreme conditions. Quantum mechanical effects tend to appear when dealing with "lows" (low temperatures, small distances), while relativistic effects tend to appear when dealing with "highs" (high velocities, large distances), the "middles" being classical behaviour. For example, when analysing the behaviour of a gas at room temperature, most phenomena will involve the (classical) Maxwell–Boltzmann distribution. However near absolute zero, the Maxwell–Boltzmann distribution fails to account for the observed behaviour of the gas, and the (modern) Fermi–Dirac or Bose–Einstein distributions have to be used instead.

German physicists Albert Einstein, founder of the theory of relativity, and Max Planck, founder of quantum theory, respectively.

Very often, it is possible to find – or "retrieve" – the classical behaviour from the modern description by analysing the modern description at low speeds and large distances (by taking a limit, or by making an approximation). When doing so, the result is called the *classical limit*.

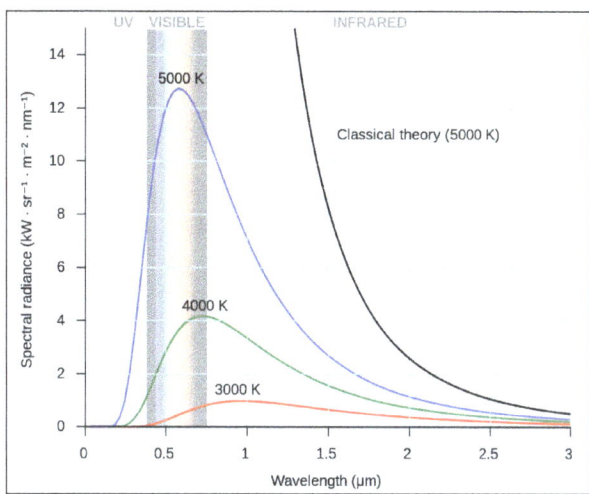

Classical physics (Rayleigh–Jeans law, black line) failed to explain black-body radiation – the so-called ultraviolet catastrophe. The quantum description (Planck's law, colored lines) is said to be modern physics.

"The term "modern physics", taken literally, means of course, the sum total of knowledge under the head of present-day physics. In this sense, the physics of 1890 is still modern; very few statements made in a good physics text of 1890 would need to be deleted today as untrue. On the other hand there have been enormous advances in physics, and some of these advances have brought into question, or have directly contradicted, certain theories that had seemed to be strongly supported by the experimental evidence.For example, few, if any physicists in 1890 questioned the wave theory of light. Its triumphs over the old corpuscular theory seemed to be final and complete, particularly after the brilliant experiments of Hertz, in 1887, which demonstrated, beyond doubt, the fundamental soundness of Maxwell's electromagnetic theory of light. And yet these very experiments of Hertz brought to light a new phenomenon—the photoelectric effect—which played an important part in establishing the quantum theory. The latter theory is diametrically opposed to the wave theory of light; indeed, the reconciliation of these two theories was one of the great problems of the first quarter of the twentieth century. "

—*F. K. Richtmyer*, E. H. Kennard, T. Lauritsen, Introduction to Modern Physics, 5th edition

Hallmarks

These are generally considered to be the topics regarded as the "core" of the foundation of modern physics:

- Atomic theory and the evolution of the atomic model in general.

- Black-body radiation.

- Oil drop experiment.

- Franck–Hertz experiment.

- Geiger–Marsden experiment (Rutherford's experiment).

- Gravitational lensing.

- Michelson–Morley experiment.

- Photoelectric effect.

- Quantum thermodynamics.

- Radioactive phenomena in general.

- Perihelion precession of Mercury.

- Stern–Gerlach experiment.

- Wave–particle duality.

- Solid-state physics.

NUCLEAR PHYSICS

Nuclear physics is the field of physics that studies atomic nuclei and their constituents and interactions. Other forms of nuclear matter are also studied. Nuclear physics should not be confused with atomic physics, which studies the atom as a whole, including its electrons.

Discoveries in nuclear physics have led to applications in many fields. This includes nuclear power, nuclear weapons, nuclear medicine and magnetic resonance imaging, industrial and agricultural isotopes, ion implantation in materials engineering, and radiocarbon dating in geology and archaeology. Such applications are studied in the field of nuclear engineering.

Particle physics evolved out of nuclear physics and the two fields are typically taught in close association. Nuclear astrophysics, the application of nuclear physics to astrophysics, is crucial in explaining the inner workings of stars and the origin of the chemical elements.

Modern Nuclear Physics

A heavy nucleus can contain hundreds of nucleons. This means that with some approximation it can be treated as a classical system, rather than a quantum-mechanical one. In the resulting liquid-drop model, the nucleus has an energy which arises partly from surface tension and partly from electrical repulsion of the protons. The liquid-drop model is able to reproduce many features of nuclei, including the general trend of binding energy with respect to mass number, as well as the phenomenon of nuclear fission.

Superimposed on this classical picture, however, are quantum-mechanical effects, which can be described using the nuclear shell model, developed in large part by Maria Goeppert Mayer and J. Hans D. Jensen. Nuclei with certain numbers of neutrons and protons are particularly stable, because their shells are filled.

Other more complicated models for the nucleus have also been proposed, such as the interacting boson model, in which pairs of neutrons and protons interact as bosons, analogously to Cooper pairs of electrons.

Ab initio methods try to solve the nuclear many-body problem from the ground up, starting from the nucleons and their interactions.

Much of current research in nuclear physics relates to the study of nuclei under extreme conditions such as high spin and excitation energy. Nuclei may also have extreme shapes (similar to that of Rugby balls or even pears) or extreme neutron-to-proton ratios. Experimenters can create such nuclei using artificially induced fusion or nucleon transfer reactions, employing ion beams from an accelerator. Beams with even higher energies can be used to create nuclei at very high temperatures, and there are signs that these experiments have produced a phase transition from normal nuclear matter to a new state, the quark–gluon plasma, in which the quarks mingle with one another, rather than being segregated in triplets as they are in neutrons and protons.

Nuclear Decay

Eighty elements have at least one stable isotope which is never observed to decay, amounting to a total of about 254 stable isotopes. However, thousands of isotopes have been characterized as unstable. These "radioisotopes" decay over time scales ranging from fractions of a second to trillions of years. Plotted on a chart as a function of atomic and neutron numbers, the binding energy of the nuclides forms what is known as the valley of stability. Stable nuclides lie along the bottom of this energy valley, while increasingly unstable nuclides lie up the valley walls, that is, have weaker binding energy.

The most stable nuclei fall within certain ranges or balances of composition of neutrons and protons: too few or too many neutrons (in relation to the number of protons) will cause it to decay. For example, in beta decay a nitrogen-16 atom (7 protons, 9 neutrons) is converted to an oxygen-16 atom (8 protons, 8 neutrons) within a few seconds of being created. In this decay a neutron in the nitrogen nucleus is converted by the weak interaction into a proton, an electron and an antineutrino. The element is transmuted to another element, with a different number of protons.

In alpha decay (which typically occurs in the heaviest nuclei) the radioactive element decays by emitting a helium nucleus (2 protons and 2 neutrons), giving another element, plus helium-4. In many cases this process continues through several steps of this kind, including other types of decays (usually beta decay) until a stable element is formed.

In gamma decay, a nucleus decays from an excited state into a lower energy state, by emitting a gamma ray. The element is not changed to another element in the process (no nuclear transmutation is involved).

Other more exotic decays are possible. For example, in internal conversion decay, the energy from an excited nucleus may eject one of the inner orbital electrons from the atom, in a process which produces high speed electrons, but is not beta decay, and (unlike beta decay) does not transmute one element to another.

Nuclear Fusion

In nuclear fusion, two low mass nuclei come into very close contact with each other, so that the strong force fuses them. It requires a large amount of energy for the strong or nuclear forces to overcome the electrical repulsion between the nuclei in order to fuse them; therefore nuclear fusion can only take place at very high temperatures or high pressures. When nuclei fuse, a very large amount of energy is released and the combined nucleus assumes a lower energy level. The binding energy per nucleon increases with mass number up to nickel-62. Stars like the Sun are powered by the fusion of four protons into a helium nucleus, two positrons, and two neutrinos. The uncontrolled fusion of hydrogen into helium is known as thermonuclear runaway. A frontier in current research at various institutions, for example the Joint European Torus (JET) and ITER, is the development of an economically viable method of using energy from a controlled fusion reaction. Nuclear fusion is the origin of the energy (including in the form of light and other electromagnetic radiation) produced by the core of all stars including our own Sun.

Nuclear Fission

Nuclear fission is the reverse process to fusion. For nuclei heavier than nickel-62 the binding energy per nucleon decreases with the mass number. It is therefore possible for energy to be released if a heavy nucleus breaks apart into two lighter ones.

The process of alpha decay is in essence a special type of spontaneous nuclear fission. It is a highly asymmetrical fission because the four particles which make up the alpha particle are especially tightly bound to each other, making production of this nucleus in fission particularly likely.

From certain of the heaviest nuclei whose fission produces free neutrons, and which also easily absorb neutrons to initiate fission, a self-igniting type of neutron-initiated fission can be obtained, in a chain reaction. Chain reactions were known in chemistry before physics, and in fact many familiar processes like fires and chemical explosions are chemical chain reactions. The fission or "nuclear" chain-reaction, using fission-produced neutrons, is the source of energy for nuclear power plants and fission type nuclear bombs, such as those detonated in Hiroshima and Nagasaki, Japan, at the end of World War II. Heavy nuclei such as uranium and thorium may also undergo spontaneous fission, but they are much more likely to undergo decay by alpha decay.

For a neutron-initiated chain reaction to occur, there must be a critical mass of the relevant isotope present in a certain space under certain conditions. The conditions for the smallest critical mass require the conservation of the emitted neutrons and also their slowing or moderation so that there is a greater cross-section or probability of them initiating another fission. In two regions of Oklo, Gabon, Africa, natural nuclear fission reactors were active over 1.5 billion years ago. Measurements of natural neutrino emission have demonstrated that around half of the heat emanating from the Earth's core results from radioactive decay. However, it is not known if any of this results from fission chain reactions.

Production of "Heavy" Elements

According to the theory, as the Universe cooled after the Big Bang it eventually became possible for common subatomic particles as we know them (neutrons, protons and electrons) to exist. The most common particles created in the Big Bang which are still easily observable to us today were protons and electrons (in equal numbers). The protons would eventually form hydrogen atoms. Almost all the neutrons created in the Big Bang were absorbed into helium-4 in the first three minutes after the Big Bang, and this helium accounts for most of the helium in the universe today.

Some relatively small quantities of elements beyond helium (lithium, beryllium, and perhaps some boron) were created in the Big Bang, as the protons and neutrons collided with each other, but all of the "heavier elements" (carbon, element number 6, and elements of greater atomic number) that we see today, were created inside stars during a series of fusion stages, such as the proton-proton chain, the CNO cycle and the triple-alpha process. Progressively heavier elements are created during the evolution of a star.

Since the binding energy per nucleon peaks around iron (56 nucleons), energy is only released in fusion processes involving smaller atoms than that. Since the creation of heavier nuclei by fusion requires energy, nature resorts to the process of neutron capture. Neutrons (due to their lack of charge) are readily absorbed by a nucleus. The heavy elements are created by either a *slow* neutron capture process (the so-called s-process) or the *rapid*, or r-process. The s process occurs in thermally pulsing stars (called AGB, or asymptotic giant branch stars) and takes hundreds to thousands of years to reach the heaviest elements of lead and bismuth. The r-process is thought to occur in supernova explosions which provide the necessary conditions of high temperature, high neutron flux and ejected matter. These stellar conditions make the successive neutron captures very fast, involving very neutron-rich species which then beta-decay to heavier elements, especially at the so-called waiting points that correspond to more stable nuclides with closed neutron shells (magic numbers).

ATOMIC PHYSICS

Atomic physics (or atom physics) is a field of physics that involves investigation of the structures of atoms, their energy states, and their interactions with other particles and electromagnetic radiation. In this field of physics, atoms are studied as isolated systems made up of nuclei and electrons. Its primary concern is related to the arrangement of electrons around the nucleus and the processes by which these arrangements change. It includes the study of atoms in the form of ions as well as in the neutral state. For purposes of this discussion, it should be assumed that the term atom includes ions, unless otherwise stated. Through studies of the structure and behavior of atoms, scientists have been able to explain and predict the properties of chemical elements, and, by extension, chemical compounds.

The term atomic physics is often associated with nuclear power and nuclear bombs, due to the synonymous use of atomic and nuclear in standard English. However, physicists distinguish between atomic physics, which deals with the atom as a system consisting of a nucleus and electrons, and nuclear physics, which considers atomic nuclei alone. As with many scientific fields, strict delineation can be highly contrived and atomic physics is often considered in the wider context of atomic, molecular, and optical physics.

Most fields of physics can be divided between theoretical work and experimental work, and atomic physics is no exception. Usually, progress alternates between experimental observations and theoretical explanations.

Clearly, the earliest steps toward atomic physics were taken with the recognition that matter is composed of atoms, in the modern sense of the basic unit of a chemical element. This theory was developed by the British chemist and physicist John Dalton in the eighteenth century. At that stage, the structures of individual atoms were not known, but atoms could be described by the properties of chemical elements, which were then organized in the form of a periodic table.

The true beginning of atomic physics was marked by the discovery of spectral lines and attempts to describe the phenomenon, most notably by Joseph von Fraunhofer. The study of these lines led to the Bohr atom model and to the birth of quantum mechanics. In seeking to explain atomic spectra, an entirely new mathematical model of matter was revealed. As far as atoms and their electron arrangements were concerned, formulation of the atomic orbital model offered a better overall description and also provided a new theoretical basis for chemistry (quantum chemistry) and spectroscopy.

Since the Second World War, both theoretical and experimental areas of atomic physics have advanced at a rapid pace. This progress can be attributed to developments in computing technology, which have allowed bigger and more sophisticated models of atomic structure and associated collision processes. Likewise, technological advances in particle accelerators, detectors, magnetic field generation, and lasers have greatly assisted experimental work in atomic physics.

Isolated Atoms

As noted above, atomic physics involves investigation of atoms as isolated entities. In atomic models, the atom is described as consisting of a single nucleus that is surrounded by one or more bound electrons. It is not concerned with the formation of molecules (although much of the physics is identical), nor does it examine atoms in a solid state as condensed matter. It is concerned with processes such as ionization and excitation by photons or collisions with atomic particles.

In practical terms, modeling atoms in isolation may not seem realistic. However, if one considers atoms in a gas or plasma, then the time scales for atom-atom interactions are huge compared to the atomic processes being examined here. This means that the individual atoms can be treated as if each were in isolation because for the vast majority of the time they are. By this consideration, atomic physics provides the underlying theory in plasma physics and atmospheric physics, although both deal with huge numbers of atoms.

Electronic Configuration

Electrons form notional shells around the nucleus. These electrons are naturally in their lowest energy state, called the ground state, but they can be excited to higher energy states by the absorption of energy from light (photons), magnetic fields, or interaction with a colliding particle (typically other electrons). The excited electron may still be bound to the nucleus, in which case they should, after a certain period of time, decay back to the original ground state. In so doing, energy is released as photons. There are strict selection rules regarding the electronic configurations that can be reached by excitation by light, but there are no such rules for excitation by collision processes.

If an electron is sufficiently excited, it may break free of the nucleus and no longer remain part of the atom. The remaining system is an ion, and the atom is said to have been ionized, having been left in a charged state.

OPTICS

Optics is the branch of physics that studies the behaviour and properties of light, including its interactions with matter and the construction of instruments that use or detect it. Optics usually describes the behaviour of visible, ultraviolet, and infrared light. Because light is an electromagnetic wave, other forms of electromagnetic radiation such as X-rays, microwaves, and radio waves exhibit similar properties.

Most optical phenomena can be accounted for using the classical electromagnetic description of light. Complete electromagnetic descriptions of light are, however, often difficult to apply in practice. Practical optics is usually done using simplified models. The most common of these, geometric optics, treats light as a collection of rays that travel in straight lines and bend when they pass through or reflect from surfaces. Physical optics is a more comprehensive model of light, which includes wave effects such as diffraction and interference that cannot be accounted for in geometric optics. Historically, the ray-based model of light was developed first, followed by the wave model of light. Progress in electromagnetic theory in the 19th century led to the discovery that light waves were in fact electromagnetic radiation.

Some phenomena depend on the fact that light has both wave-like and particle-like properties. Explanation of these effects requires quantum mechanics. When considering light's particle-like properties, the light is modelled as a collection of particles called "photons". Quantum optics deals with the application of quantum mechanics to optical systems.

Optical science is relevant to and studied in many related disciplines including astronomy, various engineering fields, photography, and medicine (particularly ophthalmology and optometry). Practical applications of optics are found in a variety of technologies and everyday objects, including mirrors, lenses, telescopes, microscopes, lasers, and fibre optics.

Classical Optics

Classical optics is divided into two main branches: geometrical (or ray) optics and physical (or wave) optics. In geometrical optics, light is considered to travel in straight lines, while in physical optics, light is considered as an electromagnetic wave.

Geometrical optics can be viewed as an approximation of physical optics that applies when the wavelength of the light used is much smaller than the size of the optical elements in the system being modelled.

Geometrical Optics

Geometrical optics, or *ray optics*, describes the propagation of light in terms of "rays" which travel in straight lines, and whose paths are governed by the laws of reflection and refraction at interfaces between different media. These laws were discovered empirically as far back as 984 AD and have been used in the design of optical components and instruments from then until the present day. They can be summarised as follows:

When a ray of light hits the boundary between two transparent materials, it is divided into a reflected and a refracted ray.

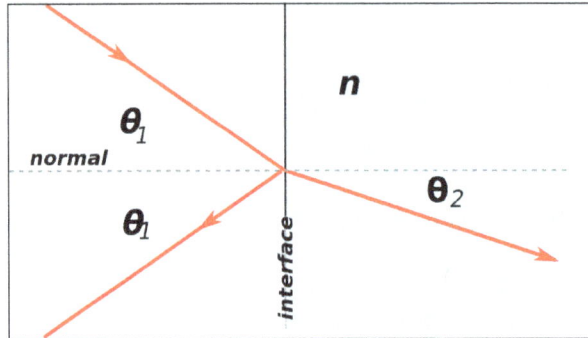

Geometry of reflection and refraction of light rays.

The law of reflection says that the reflected ray lies in the plane of incidence, and the angle of reflection equals the angle of incidence.

The law of refraction says that the refracted ray lies in the plane of incidence, and the sine of the angle of refraction divided by the sine of the angle of incidence is a constant:

$$\frac{\sin \theta_1}{\sin \theta_2} = n,$$

where n is a constant for any two materials and a given colour of light. If the first material is air or vacuum, n is the refractive index of the second material.

The laws of reflection and refraction can be derived from Fermat's principle which states that the path taken between two points by a ray of light is the path that can be traversed in the least time.

Approximations

Geometric optics is often simplified by making the paraxial approximation, or "small angle approximation". The mathematical behaviour then becomes linear, allowing optical components and systems to be described by simple matrices. This leads to the techniques of Gaussian optics and *paraxial ray tracing*, which are used to find basic properties of optical systems, such as approximate image and object positions and magnifications.

Reflections

Reflections can be divided into two types: specular reflection and diffuse reflection. Specular reflection describes the gloss of surfaces such as mirrors, which reflect light in a simple, predictable way. This allows for production of reflected images that can be associated with an actual (real) or extrapolated (virtual) location in space. Diffuse reflection describes non-glossy materials, such as paper or rock. The reflections from these surfaces can only be described statistically, with the exact distribution of the reflected light depending on the microscopic structure of the material. Many diffuse reflectors are described or can be approximated by Lambert's cosine law, which describes surfaces that have equal luminance when viewed from any angle. Glossy surfaces can give both specular and diffuse reflection.

In specular reflection, the direction of the reflected ray is determined by the angle the incident ray makes with the surface normal, a line perpendicular to the surface at the point where the ray hits.

The incident and reflected rays and the normal lie in a single plane, and the angle between the reflected ray and the surface normal is the same as that between the incident ray and the normal. This is known as the Law of Reflection.

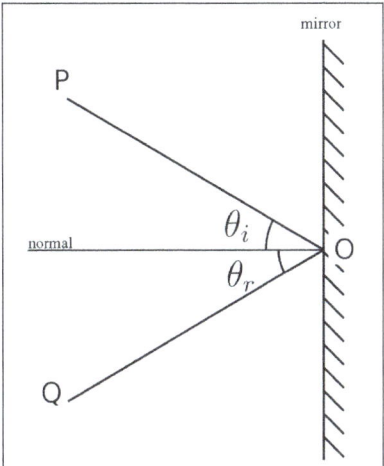

Diagram of specular reflection.

For flat mirrors, the law of reflection implies that images of objects are upright and the same distance behind the mirror as the objects are in front of the mirror. The image size is the same as the object size. The law also implies that mirror images are parity inverted, which we perceive as a left-right inversion. Images formed from reflection in two (or any even number of) mirrors are not parity inverted. Corner reflectors retroreflect light, producing reflected rays that travel back in the direction from which the incident rays came. This is called retroreflection.

Mirrors with curved surfaces can be modelled by ray tracing and using the law of reflection at each point on the surface. For mirrors with parabolic surfaces, parallel rays incident on the mirror produce reflected rays that converge at a common focus. Other curved surfaces may also focus light, but with aberrations due to the diverging shape causing the focus to be smeared out in space. In particular, spherical mirrors exhibit spherical aberration. Curved mirrors can form images with magnification greater than or less than one, and the magnification can be negative, indicating that the image is inverted. An upright image formed by reflection in a mirror is always virtual, while an inverted image is real and can be projected onto a screen.

Refractions

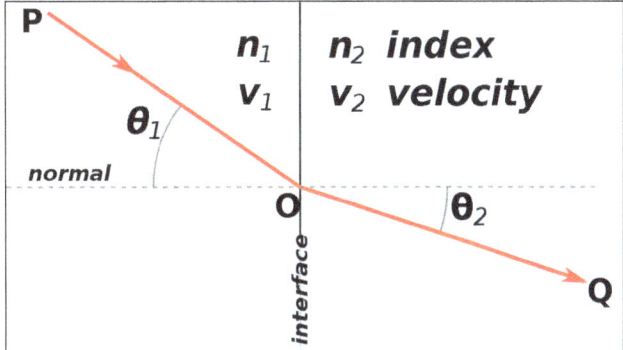

Illustration of Snell's Law for the case n1 < n2, such as air/water interface.

Refraction occurs when light travels through an area of space that has a changing index of refraction; this principle allows for lenses and the focusing of light. The simplest case of refraction occurs when there is an interface between a uniform medium with index of refraction n_1 and another medium with index of refraction n_2. In such situations, Snell's Law describes the resulting deflection of the light ray:

$$n_1 \sin \theta_1 = n_2 \sin \theta_2$$

where θ_1 and θ_2 are the angles between the normal (to the interface) and the incident and refracted waves, respectively.

The index of refraction of a medium is related to the speed, v, of light in that medium by

$$n = c / v,$$

where c is the speed of light in vacuum.

Snell's Law can be used to predict the deflection of light rays as they pass through linear media as long as the indexes of refraction and the geometry of the media are known. For example, the propagation of light through a prism results in the light ray being deflected depending on the shape and orientation of the prism. In most materials, the index of refraction varies with the frequency of the light. Taking this into account, Snell's Law can be used to predict how a prism will disperse light into a spectrum. The discovery of this phenomenon when passing light through a prism is famously attributed to Isaac Newton.

Some media have an index of refraction which varies gradually with position and, therefore, light rays in the medium are curved. This effect is responsible for mirages seen on hot days: a change in index of refraction air with height causes light rays to bend, creating the appearance of specular reflections in the distance (as if on the surface of a pool of water). Optical materials with varying index of refraction are called gradient-index (GRIN) materials. Such materials are used to make gradient-index optics.

For light rays travelling from a material with a high index of refraction to a material with a low index of refraction, Snell's law predicts that there is no θ_2 when θ_1 is large. In this case, no transmission occurs; all the light is reflected. This phenomenon is called total internal reflection and allows for fibre optics technology. As light travels down an optical fibre, it undergoes total internal reflection allowing for essentially no light to be lost over the length of the cable.

Lenses

A device which produces converging or diverging light rays due to refraction is known as a *lens*. Lenses are characterized by their focal length: a converging lens has positive focal length, while a diverging lens has negative focal length. Smaller focal length indicates that the lens has a stronger converging or diverging effect. The focal length of a simple lens in air is given by the lensmaker's equation.

Ray tracing can be used to show how images are formed by a lens. For a thin lens in air, the location of the image is given by the simple equation

$$\frac{1}{S_1} + \frac{1}{S_2} = \frac{1}{f},$$

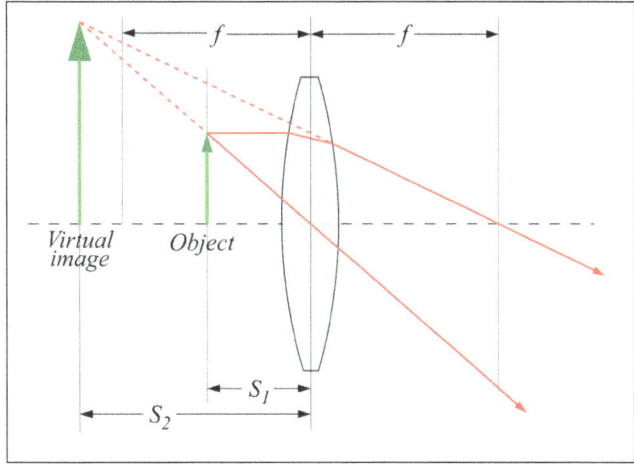

A ray tracing diagram for a converging lens.

where S_1 is the distance from the object to the lens, S_2 is the distance from the lens to the image, and f is the focal length of the lens. In the sign convention used here, the object and image distances are positive if the object and image are on opposite sides of the lens.

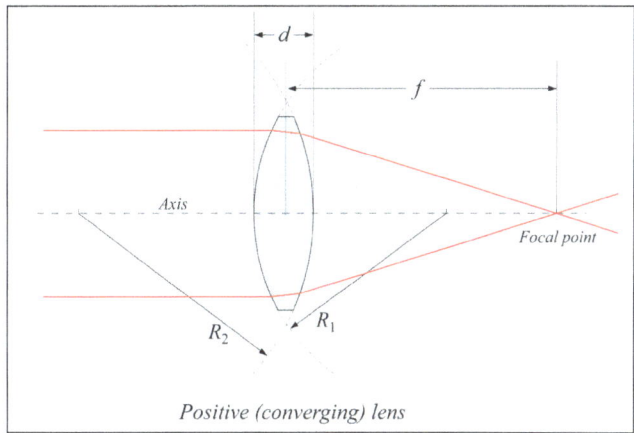

Positive (converging) lens

Incoming parallel rays are focused by a converging lens onto a spot one focal length from the lens, on the far side of the lens. This is called the rear focal point of the lens. Rays from an object at finite distance are focused further from the lens than the focal distance; the closer the object is to the lens, the further the image is from the lens.

With diverging lenses, incoming parallel rays diverge after going through the lens, in such a way that they seem to have originated at a spot one focal length in front of the lens. This is the lens's front focal point. Rays from an object at finite distance are associated with a virtual image that is closer to the lens than the focal point, and on the same side of the lens as the object. The closer the object is to the lens, the closer the virtual image is to the lens. As with mirrors, upright images produced by a single lens are virtual, while inverted images are real.

Lenses suffer from aberrations that distort images. *Monochromatic aberrations* occur because the geometry of the lens does not perfectly direct rays from each object point to a single point on the image, while chromatic aberration occurs because the index of refraction of the lens varies with the wavelength of the light.

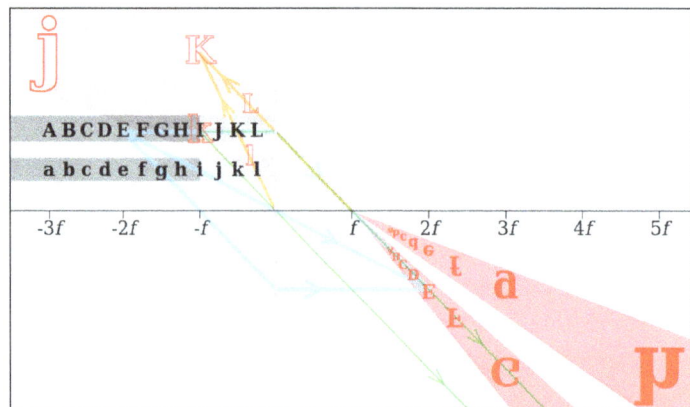

Images of black letters in a thin convex lens of focal length f are shown in red. Selected rays are shown for letters E, I and K in blue, green and orange, respectively. Note that E (at 2f) has an equal-size, real and inverted image; I (at f) has its image at infinity; and K (at f/2) has a double-size, virtual and upright image.

Physical Optics

In physical optics, light is considered to propagate as a wave. This model predicts phenomena such as interference and diffraction, which are not explained by geometric optics. The speed of light waves in air is approximately 3.0×10^8 m/s (exactly 299,792,458 m/s in vacuum). The wavelength of visible light waves varies between 400 and 700 nm, but the term "light" is also often applied to infrared (0.7–300 µm) and ultraviolet radiation (10–400 nm).

The wave model can be used to make predictions about how an optical system will behave without requiring an explanation of what is "waving" in what medium. Until the middle of the 19th century, most physicists believed in an "ethereal" medium in which the light disturbance propagated. The existence of electromagnetic waves was predicted in 1865 by Maxwell's equations. These waves propagate at the speed of light and have varying electric and magnetic fields which are orthogonal to one another, and also to the direction of propagation of the waves. Light waves are now generally treated as electromagnetic waves except when quantum mechanical effects have to be considered.

Modelling and Design of Optical Systems using Physical Optics

Many simplified approximations are available for analysing and designing optical systems. Most of these use a single scalar quantity to represent the electric field of the light wave, rather than using a vector model with orthogonal electric and magnetic vectors. The Huygens–Fresnel equation is one such model. This was derived empirically by Fresnel in 1815, based on Huygens' hypothesis that each point on a wavefront generates a secondary spherical wavefront, which Fresnel combined with the principle of superposition of waves. The Kirchhoff diffraction equation, which is derived using Maxwell's equations, puts the Huygens-Fresnel equation on a firmer physical foundation. Examples of the application of Huygens–Fresnel principle can be found in the sections on diffraction and Fraunhofer diffraction.

More rigorous models, involving the modelling of both electric and magnetic fields of the light wave, are required when dealing with the detailed interaction of light with materials where the interaction depends on their electric and magnetic properties. For instance, the behaviour of a light wave interacting with a metal surface is quite different from what happens when it interacts with a dielectric material. A vector model must also be used to model polarised light.

Numerical modeling techniques such as the finite element method, the boundary element method and the transmission-line matrix method can be used to model the propagation of light in systems which cannot be solved analytically. Such models are computationally demanding and are normally only used to solve small-scale problems that require accuracy beyond that which can be achieved with analytical solutions.

All of the results from geometrical optics can be recovered using the techniques of Fourier optics which apply many of the same mathematical and analytical techniques used in acoustic engineering and signal processing.

Gaussian beam propagation is a simple paraxial physical optics model for the propagation of coherent radiation such as laser beams. This technique partially accounts for diffraction, allowing accurate calculations of the rate at which a laser beam expands with distance, and the minimum size to which the beam can be focused. Gaussian beam propagation thus bridges the gap between geometric and physical optics.

Superposition and Interference

In the absence of nonlinear effects, the superposition principle can be used to predict the shape of interacting waveforms through the simple addition of the disturbances. This interaction of waves to produce a resulting pattern is generally termed "interference" and can result in a variety of outcomes. If two waves of the same wavelength and frequency are *in phase*, both the wave crests and wave troughs align. This results in constructive interference and an increase in the amplitude of the wave, which for light is associated with a brightening of the waveform in that location. Alternatively, if the two waves of the same wavelength and frequency are out of phase, then the wave crests will align with wave troughs and vice versa. This results in destructive interference and a decrease in the amplitude of the wave, which for light is associated with a dimming of the waveform at that location.

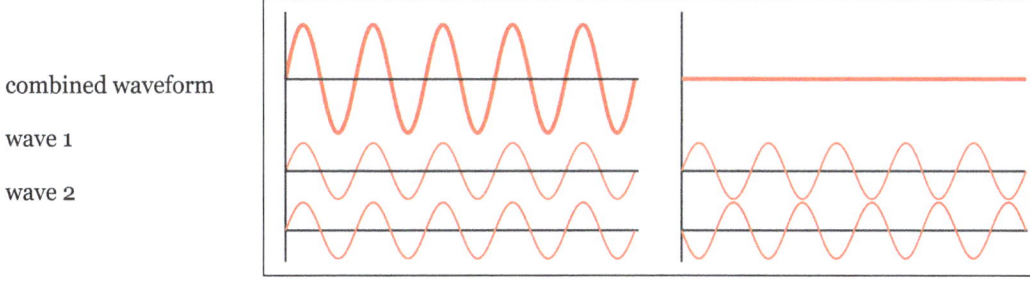

combined waveform

wave 1

wave 2

Two waves in phase.   Two waves 180° out of phase.

Since the Huygens–Fresnel principle states that every point of a wavefront is associated with the production of a new disturbance, it is possible for a wavefront to interfere with itself constructively or destructively at different locations producing bright and dark fringes in regular and predictable patterns. Interferometry is the science of measuring these patterns, usually as a means of making precise determinations of distances or angular resolutions. The Michelson interferometer was a famous instrument which used interference effects to accurately measure the speed of light.

When oil or fuel is spilled, colourful patterns are formed by thin-film interference.

The appearance of thin films and coatings is directly affected by interference effects. Antireflective coatings use destructive interference to reduce the reflectivity of the surfaces they coat, and can be used to minimise glare and unwanted reflections. The simplest case is a single layer with thickness one-fourth the wavelength of incident light. The reflected wave from the top of the film and the reflected wave from the film/material interface are then exactly 180° out of phase, causing destructive interference. The waves are only exactly out of phase for one wavelength, which would typically be chosen to be near the centre of the visible spectrum, around 550 nm. More complex designs using multiple layers can achieve low reflectivity over a broad band, or extremely low reflectivity at a single wavelength.

Constructive interference in thin films can create strong reflection of light in a range of wavelengths, which can be narrow or broad depending on the design of the coating. These films are used to make dielectric mirrors, interference filters, heat reflectors, and filters for colour separation in colour television cameras. This interference effect is also what causes the colourful rainbow patterns seen in oil slicks.

Diffraction and Optical Resolution

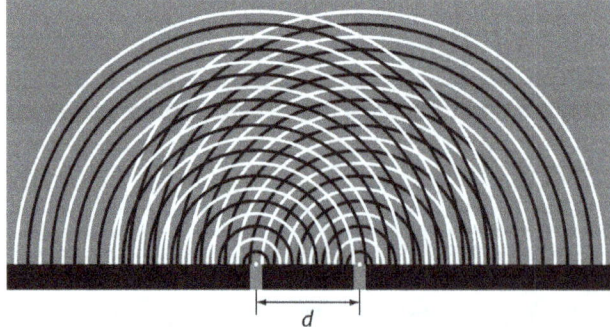

Diffraction on two slits separated by distance d. The bright fringes occur along lines where black lines intersect with black lines and white lines intersect with white lines. These fringes are separated by angle θ and are numbered as order n.

Diffraction is the process by which light interference is most commonly observed. The effect was first described in 1665 by Francesco Maria Grimaldi, who also coined the term from the Latin *diffringere*, 'to break into pieces'. Later that century, Robert Hooke and Isaac Newton also described phenomena now known to be diffraction in Newton's rings while James Gregory recorded his observations of diffraction patterns from bird feathers.

The first physical optics model of diffraction that relied on the Huygens–Fresnel principle was developed in 1803 by Thomas Young in his interference experiments with the interference patterns of two closely spaced slits. Young showed that his results could only be explained if the two slits acted as two unique sources of waves rather than corpuscles. In 1815 and 1818, Augustin-Jean Fresnel firmly established the mathematics of how wave interference can account for diffraction.

The simplest physical models of diffraction use equations that describe the angular separation of light and dark fringes due to light of a particular wavelength (λ). In general, the equation takes the form

$$m\lambda = d \sin \theta$$

where d is the separation between two wavefront sources (in the case of Young's experiments, it was two slits), θ is the angular separation between the central fringe and the m th order fringe, where the central maximum is $m = 0$.

This equation is modified slightly to take into account a variety of situations such as diffraction through a single gap, diffraction through multiple slits, or diffraction through a diffraction grating that contains a large number of slits at equal spacing. More complicated models of diffraction require working with the mathematics of Fresnel or Fraunhofer diffraction.

X-ray diffraction makes use of the fact that atoms in a crystal have regular spacing at distances that are on the order of one angstrom. To see diffraction patterns, x-rays with similar wavelengths to that spacing are passed through the crystal. Since crystals are three-dimensional objects rather than two-dimensional gratings, the associated diffraction pattern varies in two directions according to Bragg reflection, with the associated bright spots occurring in unique patterns and d being twice the spacing between atoms.

Diffraction effects limit the ability for an optical detector to optically resolve separate light sources. In general, light that is passing through an aperture will experience diffraction and the best images that can be created (as described in diffraction-limited optics) appear as a central spot with surrounding bright rings, separated by dark nulls; this pattern is known as an Airy pattern, and the central bright lobe as an Airy disk. The size of such a disk is given by

$$\sin \theta = 1.22 \frac{\lambda}{D}$$

where θ is the angular resolution, λ is the wavelength of the light, and D is the diameter of the lens aperture. If the angular separation of the two points is significantly less than the Airy disk angular radius, then the two points cannot be resolved in the image, but if their angular separation is much greater than this, distinct images of the two points are formed and they can therefore be resolved. Rayleigh defined the somewhat arbitrary "Rayleigh criterion" that two points whose angular separation is equal to the Airy disk radius (measured to first null, that is, to the first place where no light is seen) can be considered to be resolved. It can be seen that the greater the diameter of the lens or its aperture, the finer the resolution. Interferometry, with its ability to mimic extremely large baseline apertures, allows for the greatest angular resolution possible.

For astronomical imaging, the atmosphere prevents optimal resolution from being achieved in the visible spectrum due to the atmospheric scattering and dispersion which cause stars to twinkle. Astronomers refer to this effect as the quality of astronomical seeing. Techniques known as adaptive optics have been used to eliminate the atmospheric disruption of images and achieve results that approach the diffraction limit.

Dispersion and Scattering

Refractive processes take place in the physical optics limit, where the wavelength of light is similar to other distances, as a kind of scattering. The simplest type of scattering is Thomson scattering which occurs when electromagnetic waves are deflected by single particles. In the limit of Thomson scattering, in which the wavelike nature of light is evident, light is dispersed independent of the frequency, in contrast to Compton scattering which is frequency-dependent and strictly a quantum mechanical process, involving the nature of light as particles. In a statistical sense, elastic scattering of light by numerous particles much smaller than the wavelength of the light is a process known as Rayleigh scattering while the similar process for scattering by particles that are similar or larger in wavelength is known as Mie scattering with the Tyndall effect being a commonly observed result. A small proportion of light scattering from atoms or molecules may undergo Raman scattering, wherein the frequency changes due to excitation of the atoms and molecules. Brillouin scattering occurs when the frequency of light changes due to local changes with time and movements of a dense material.

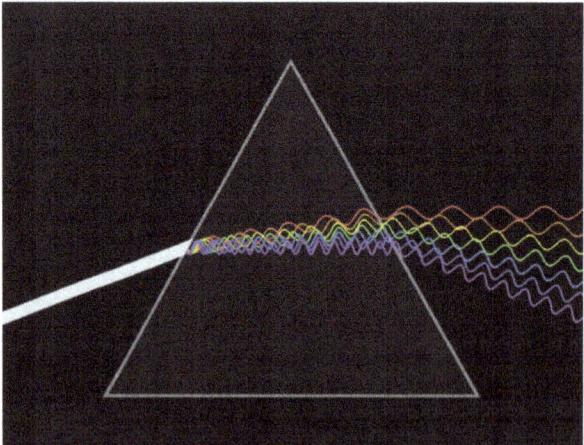

Conceptual animation of light dispersion through a prism. High frequency (blue)
light is deflected the most, and low frequency (red) the least.

Dispersion occurs when different frequencies of light have different phase velocities, due either to material properties (*material dispersion*) or to the geometry of an optical waveguide (*waveguide dispersion*). The most familiar form of dispersion is a decrease in index of refraction with increasing wavelength, which is seen in most transparent materials. This is called "normal dispersion". It occurs in all dielectric materials, in wavelength ranges where the material does not absorb light. In wavelength ranges where a medium has significant absorption, the index of refraction can increase with wavelength. This is called "anomalous dispersion".

The separation of colours by a prism is an example of normal dispersion. At the surfaces of the prism, Snell's law predicts that light incident at an angle θ to the normal will be refracted at an angle arcsin(sin (θ) / n). Thus, blue light, with its higher refractive index, is bent more strongly than red light, resulting in the well-known rainbow pattern.

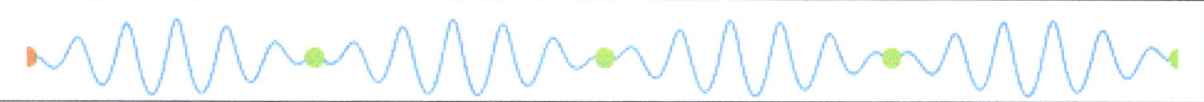

Dispersion: two sinusoids propagating at different speeds make a moving interference pattern. The red dot moves with the phase velocity, and the green dots propagate with the group velocity. In this case, the phase velocity is twice the group velocity. The red dot overtakes two green dots, when moving from the left to the right of the figure. In effect, the individual waves (which travel with the phase velocity) escape from the wave packet (which travels with the group velocity).

Material dispersion is often characterised by the Abbe number, which gives a simple measure of dispersion based on the index of refraction at three specific wavelengths. Waveguide dispersion is dependent on the propagation constant. Both kinds of dispersion cause changes in the group characteristics of the wave, the features of the wave packet that change with the same frequency as the amplitude of the electromagnetic wave. "Group velocity dispersion" manifests as a spreading-out of the signal "envelope" of the radiation and can be quantified with a group dispersion delay parameter:

$$D = \frac{1}{v_g^2} \frac{dv_g}{d\lambda}$$

where v_g is the group velocity. For a uniform medium, the group velocity is

$$v_g = c \left(n - \lambda \frac{dn}{d\lambda} \right)^{-1}$$

where n is the index of refraction and c is the speed of light in a vacuum. This gives a simpler form for the dispersion delay parameter:

$$D = -\frac{\lambda}{c} \frac{d^2 n}{d\lambda^2}.$$

If D is less than zero, the medium is said to have *positive dispersion* or normal dispersion. If D is greater than zero, the medium has *negative dispersion*. If a light pulse is propagated through a normally dispersive medium, the result is the higher frequency components slow down more than the lower frequency components. The pulse therefore becomes *positively chirped*, or *up-chirped*, increasing in frequency with time. This causes the spectrum coming out of a prism to appear with red light the least refracted and blue/violet light the most refracted. Conversely, if a pulse travels through an anomalously (negatively) dispersive medium, high frequency components travel faster than the lower ones, and the pulse becomes *negatively chirped*, or *down-chirped*, decreasing in frequency with time.

The result of group velocity dispersion, whether negative or positive, is ultimately temporal spreading of the pulse. This makes dispersion management extremely important in optical communications systems based on optical fibres, since if dispersion is too high, a group of pulses representing information will each spread in time and merge, making it impossible to extract the signal.

Polarization

Polarization is a general property of waves that describes the orientation of their oscillations. For transverse waves such as many electromagnetic waves, it describes the orientation of the oscillations in the plane perpendicular to the wave's direction of travel. The oscillations may be oriented in a single direction (linear polarization), or the oscillation direction may rotate as the wave travels (circular or elliptical polarization). Circularly polarised waves can rotate rightward or leftward in the direction of travel, and which of those two rotations is present in a wave is called the wave's chirality.

The typical way to consider polarization is to keep track of the orientation of the electric field vector as the electromagnetic wave propagates. The electric field vector of a plane wave may be arbitrarily divided into two perpendicular components labeled x and y (with $\mathbf{z}$ indicating the direction of travel). The shape traced out in the x-y plane by the electric field vector is a Lissajous figure that describes the *polarization state*. The following figures show some examples of the evolution of the electric field vector (blue), with time (the vertical axes), at a particular point in space, along with its x and y components (red/left and green/right), and the path traced by the vector in the plane (purple): The same evolution would occur when looking at the electric field at a particular time while evolving the point in space, along the direction opposite to propagation.

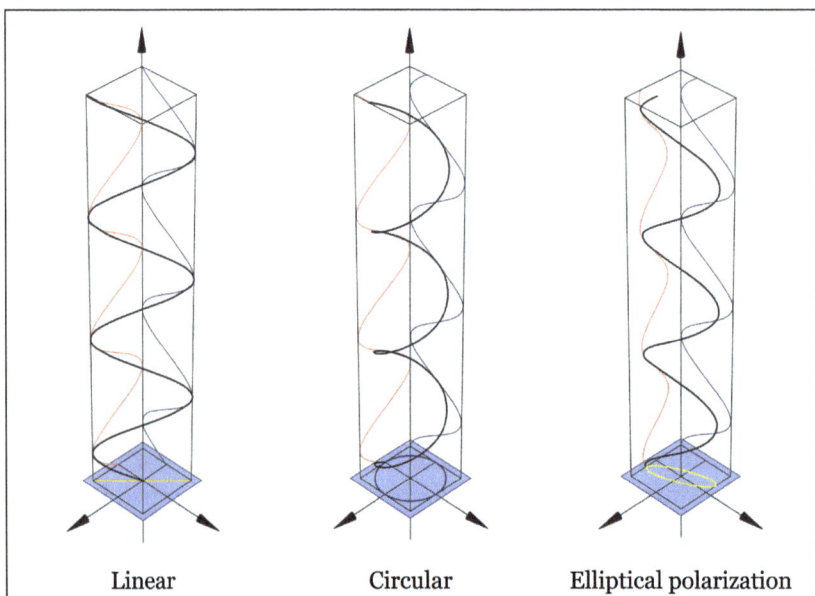

| Linear | Circular | Elliptical polarization |

In the leftmost figure above, the x and y components of the light wave are in phase. In this case, the ratio of their strengths is constant, so the direction of the electric vector (the vector sum of these two components) is constant. Since the tip of the vector traces out a single line in the plane, this special case is called linear polarization. The direction of this line depends on the relative amplitudes of the two components.

In the middle figure, the two orthogonal components have the same amplitudes and are 90° out of phase. In this case, one component is zero when the other component is at maximum or minimum amplitude. There are two possible phase relationships that satisfy this requirement: the x component can be 90° ahead of the y component or it can be 90° behind the y component. In this special case, the electric vector traces out a circle in the plane, so this polarization is called circular polarization. The rotation direction in the circle depends on which of the two phase relationships exists and corresponds to *right-hand circular polarization* and *left-hand circular polarization*.

In all other cases, where the two components either do not have the same amplitudes and/or their phase difference is neither zero nor a multiple of 90°, the polarization is called elliptical polarization because the electric vector traces out an ellipse in the plane (the *polarization ellipse*). This is shown in the above figure on the right. Detailed mathematics of polarization is done using Jones calculus and is characterised by the Stokes parameters.

Changing Polarization

Media that have different indexes of refraction for different polarization modes are called *birefringent*. Well known manifestations of this effect appear in optical wave plates/retarders (linear modes) and in Faraday rotation/optical rotation (circular modes). If the path length in the birefringent medium is sufficient, plane waves will exit the material with a significantly different propagation direction, due to refraction. For example, this is the case with macroscopic crystals of calcite, which present the viewer with two offset, orthogonally polarised images of whatever is viewed through them. It was this effect that provided the first discovery of polarization, by Erasmus Bartholinus in 1669. In addition, the phase shift, and thus the change in polarization state, is usually frequency dependent, which, in combination with dichroism, often gives rise to bright colours and rainbow-like effects. In mineralogy, such properties, known as pleochroism, are frequently exploited for the purpose of identifying minerals using polarization microscopes. Additionally, many plastics that are not normally birefringent will become so when subject to mechanical stress, a phenomenon which is the basis of photoelasticity. Non-birefringent methods, to rotate the linear polarization of light beams, include the use of prismatic polarization rotators which use total internal reflection in a prism set designed for efficient collinear transmission.

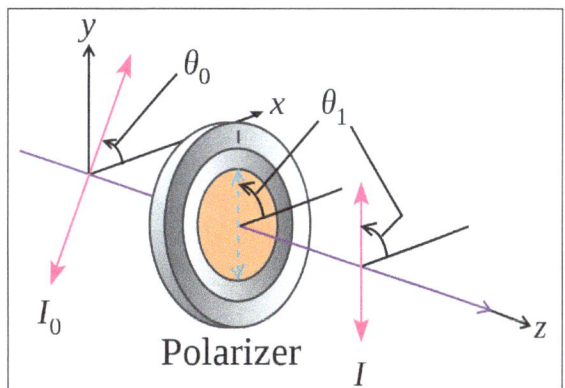

A polariser changing the orientation of linearly polarised light.

In this picture, $\theta_1 - \theta_o = \theta_i$.

Media that reduce the amplitude of certain polarization modes are called *dichroic*, with devices that block nearly all of the radiation in one mode known as *polarizing filters* or simply "polarisers". Malus' law, which is named after Étienne-Louis Malus, says that when a perfect polariser is placed in a linear polarised beam of light, the intensity, I, of the light that passes through is given by

$$I = I_0 \cos^2 \theta_i \quad ,$$

where,

I_0 is the initial intensity,

and θ_i is the angle between the light's initial polarization direction and the axis of the polariser.

A beam of unpolarised light can be thought of as containing a uniform mixture of linear polarizations at all possible angles. Since the average value of $\cos^2 \theta$ is 1/2, the transmission coefficient becomes

$$\frac{I}{I_0} = \frac{1}{2}$$

In practice, some light is lost in the polariser and the actual transmission of unpolarised light will be somewhat lower than this, around 38% for Polaroid-type polarisers but considerably higher (>49.9%) for some birefringent prism types.

In addition to birefringence and dichroism in extended media, polarization effects can also occur at the (reflective) interface between two materials of different refractive index. These effects are treated by the Fresnel equations. Part of the wave is transmitted and part is reflected, with the ratio depending on angle of incidence and the angle of refraction. In this way, physical optics recovers Brewster's angle. When light reflects from a thin film on a surface, interference between the reflections from the film's surfaces can produce polarization in the reflected and transmitted light.

Natural Light

The effects of a polarising filter on the sky in a photograph. Left picture is taken without polariser. For the right picture, filter was adjusted to eliminate certain polarizations of the scattered blue light from the sky.

Most sources of electromagnetic radiation contain a large number of atoms or molecules that emit light. The orientation of the electric fields produced by these emitters may not be correlated, in which case the light is said to be *unpolarised*. If there is partial correlation between the emitters, the light is *partially polarised*. If the polarization is consistent across the spectrum of the source, partially polarised light can be described as a superposition of a completely unpolarised component, and a completely polarised one. One may then describe the light in terms of the degree of polarization, and the parameters of the polarization ellipse.

Light reflected by shiny transparent materials is partly or fully polarised, except when the light is normal (perpendicular) to the surface. It was this effect that allowed the mathematician Étienne-Louis Malus to make the measurements that allowed for his development of the first mathematical models for polarised light. Polarization occurs when light is scattered in the atmosphere. The scattered light produces the brightness and colour in clear skies. This partial polarization of scattered light can be taken advantage of using polarizing filters to darken the sky in photographs. Optical polarization is principally of importance in chemistry due to circular dichroism and optical rotation ("*circular birefringence*") exhibited by optically active (chiral) molecules.

Modern Optics

Modern optics encompasses the areas of optical science and engineering that became popular in the 20th century. These areas of optical science typically relate to the electromagnetic or quantum properties of light but do include other topics. A major subfield of modern optics, quantum optics, deals with specifically quantum mechanical properties of light. Quantum optics is not just theoretical; some modern devices, such as lasers, have principles of operation that depend on quantum mechanics. Light detectors, such as photomultipliers and channeltrons, respond to individual photons. Electronic image sensors, such as CCDs, exhibit shot noise corresponding to the statistics of individual photon events. Light-emitting diodes and photovoltaic cells, too, cannot be understood without quantum mechanics. In the study of these devices, quantum optics often overlaps with quantum electronics.

Specialty areas of optics research include the study of how light interacts with specific materials as in crystal optics and metamaterials. Other research focuses on the phenomenology of electromagnetic waves as in singular optics, non-imaging optics, non-linear optics, statistical optics, and radiometry. Additionally, computer engineers have taken an interest in integrated optics, machine vision, and photonic computing as possible components of the "next generation" of computers.

Today, the pure science of optics is called optical science or optical physics to distinguish it from applied optical sciences, which are referred to as optical engineering. Prominent subfields of optical engineering include illumination engineering, photonics, and optoelectronics with practical applications like lens design, fabrication and testing of optical components, and image processing. Some of these fields overlap, with nebulous boundaries between the subjects terms that mean slightly different things in different parts of the world and in different areas of industry. A professional community of researchers in nonlinear optics has developed in the last several decades due to advances in laser technology.

Lasers

A laser is a device that emits light (electromagnetic radiation) through a process called *stimulated emission*. The term *laser* is an acronym for *Light Amplification by Stimulated Emission of Radiation*. Laser light is usually spatially coherent, which means that the light either is emitted in a narrow, low-divergence beam, or can be converted into one with the help of optical components such as lenses. Because the microwave equivalent of the laser, the *maser*, was developed first, devices that emit microwave and radio frequencies are usually called *masers*.

Experiments such as this one with high-power lasers are part of the modern optics research.

VLT's laser guided star.

The first working laser was demonstrated on 16 May 1960 by Theodore Maiman at Hughes Research Laboratories. When first invented, they were called "a solution looking for a problem". Since then, lasers have become a multibillion-dollar industry, finding utility in thousands of highly varied applications. The first application of lasers visible in the daily lives of the general population was the supermarket barcode scanner, introduced in 1974. The laserdisc player, introduced in 1978, was the first successful consumer product to include a laser, but the compact disc player was the first laser-equipped device to become truly common in consumers' homes, beginning in 1982. These optical storage devices use a semiconductor laser less than a millimetre wide to scan the surface of the disc for data retrieval. Fibre-optic communication relies on lasers to transmit large amounts of information at the speed of light. Other common applications of lasers include laser printers and laser pointers. Lasers are used in medicine in areas such as bloodless surgery, laser eye surgery, and laser capture microdissection and in military applications such as missile defence systems, electro-optical countermeasures (EOCM), and lidar. Lasers are also used in holograms, bubblegrams, laser light shows, and laser hair removal.

Kapitsa–Dirac Effect

The Kapitsa–Dirac effect causes beams of particles to diffract as the result of meeting a standing wave of light. Light can be used to position matter using various phenomena.

Applications

Optics is part of everyday life. The ubiquity of visual systems in biology indicates the central role optics plays as the science of one of the five senses. Many people benefit from eyeglasses or contact lenses, and optics are integral to the functioning of many consumer goods including cameras. Rainbows and mirages are examples of optical phenomena. Optical communication provides the backbone for both the Internet and modern telephony.

Human Eye

The human eye functions by focusing light onto a layer of photoreceptor cells called the retina, which forms the inner lining of the back of the eye. The focusing is accomplished by a series of transparent media. Light entering the eye passes first through the cornea, which provides much of the eye's optical power. The light then continues through the fluid just behind the cornea—the anterior chamber, then passes through the pupil. The light then passes through the lens, which focuses the light further and allows adjustment of focus. The light then passes through the main body of fluid in the eye—the vitreous humour, and reaches the retina. The cells in the retina line the back of the eye, except for where the optic nerve exits; this results in a blind spot.

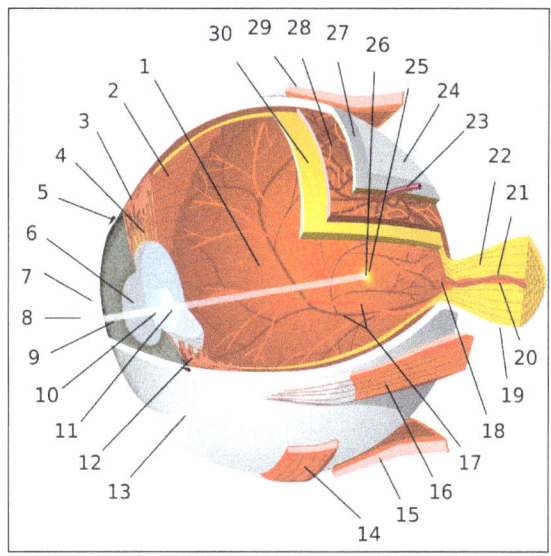

Model of a human eye. Features mentioned are 3. ciliary muscle, 6. pupil, 8. cornea, 10. lens cortex, 22. optic nerve, 26. fovea, 30. retina.

There are two types of photoreceptor cells, rods and cones, which are sensitive to different aspects of light. Rod cells are sensitive to the intensity of light over a wide frequency range, thus are responsible for black-and-white vision. Rod cells are not present on the fovea, the area of the retina responsible for central vision, and are not as responsive as cone cells to spatial and temporal changes in light. There are, however, twenty times more rod cells than cone cells in the retina because the rod cells are present across a wider area. Because of their wider distribution, rods are responsible for peripheral vision.

In contrast, cone cells are less sensitive to the overall intensity of light, but come in three varieties that are sensitive to different frequency-ranges and thus are used in the perception of colour and photopic vision. Cone cells are highly concentrated in the fovea and have a high visual acuity meaning that they are better at spatial resolution than rod cells. Since cone cells are not as sensitive to dim light as rod cells, most night vision is limited to rod cells. Likewise, since cone cells are in the fovea, central vision (including the vision needed to do most reading, fine detail work such as sewing, or careful examination of objects) is done by cone cells.

Ciliary muscles around the lens allow the eye's focus to be adjusted. This process is known as accommodation. The near point and far point define the nearest and farthest distances from the eye at which an object can be brought into sharp focus. For a person with normal vision, the far point is located at infinity. The near point's location depends on how much the muscles can increase the curvature of the lens, and how inflexible the lens has become with age. Optometrists, ophthalmologists, and opticians usually consider an appropriate near point to be closer than normal reading distance—approximately 25 cm.

Defects in vision can be explained using optical principles. As people age, the lens becomes less flexible and the near point recedes from the eye, a condition known as presbyopia. Similarly, people suffering from hyperopia cannot decrease the focal length of their lens enough to allow for nearby objects to be imaged on their retina. Conversely, people who cannot increase the focal length of their lens enough to allow for distant objects to be imaged on the retina suffer from myopia and have a far point that is considerably closer than infinity. A condition known as astigmatism results when the cornea is not spherical but instead is more curved in one direction. This causes horizontally extended objects to be focused on different parts of the retina than vertically extended objects, and results in distorted images.

All of these conditions can be corrected using corrective lenses. For presbyopia and hyperopia, a converging lens provides the extra curvature necessary to bring the near point closer to the eye while for myopia a diverging lens provides the curvature necessary to send the far point to infinity. Astigmatism is corrected with a cylindrical surface lens that curves more strongly in one direction than in another, compensating for the non-uniformity of the cornea.

The optical power of corrective lenses is measured in diopters, a value equal to the reciprocal of the focal length measured in metres; with a positive focal length corresponding to a converging lens and a negative focal length corresponding to a diverging lens. For lenses that correct for astigmatism as well, three numbers are given: one for the spherical power, one for the cylindrical power, and one for the angle of orientation of the astigmatism.

Visual Effects

The Ponzo Illusion relies on the fact that parallel lines appear to converge as they approach infinity.

Optical illusions (also called visual illusions) are characterized by visually perceived images that differ from objective reality. The information gathered by the eye is processed in the brain to give a percept that differs from the object being imaged. Optical illusions can be the result of a variety of phenomena including physical effects that create images that are different from the objects that make them, the physiological effects on the eyes and brain of excessive stimulation (e.g. brightness, tilt, colour, movement), and cognitive illusions where the eye and brain make unconscious inferences.

Cognitive illusions include some which result from the unconscious misapplication of certain optical principles. For example, the Ames room, Hering, Müller-Lyer, Orbison, Ponzo, Sander, and Wundt illusions all rely on the suggestion of the appearance of distance by using converging and diverging lines, in the same way that parallel light rays (or indeed any set of parallel lines) appear to converge at a vanishing point at infinity in two-dimensionally rendered images with artistic perspective. This suggestion is also responsible for the famous moon illusion where the moon, despite having essentially the same angular size, appears much larger near the horizon than it does at zenith. This illusion so confounded Ptolemy that he incorrectly attributed it to atmospheric refraction when he described it in his treatise, *Optics*.

Another type of optical illusion exploits broken patterns to trick the mind into perceiving symmetries or asymmetries that are not present. Examples include the café wall, Ehrenstein, Fraser spiral, Poggendorff, and Zöllner illusions. Related, but not strictly illusions, are patterns that occur due to the superimposition of periodic structures. For example, transparent tissues with a grid structure produce shapes known as moiré patterns, while the superimposition of periodic transparent patterns comprising parallel opaque lines or curves produces line moiré patterns.

Optical Instruments

Single lenses have a variety of applications including photographic lenses, corrective lenses, and magnifying glasses while single mirrors are used in parabolic reflectors and rear-view mirrors. Combining a number of mirrors, prisms, and lenses produces compound optical instruments which have practical uses. For example, a periscope is simply two plane mirrors aligned to allow for viewing around obstructions. The most famous compound optical instruments in science are the microscope and the telescope which were both invented by the Dutch in the late 16th century.

Microscopes were first developed with just two lenses: an objective lens and an eyepiece. The objective lens is essentially a magnifying glass and was designed with a very small focal length while the eyepiece generally has a longer focal length. This has the effect of producing magnified images of close objects. Generally, an additional source of illumination is used since magnified images are dimmer due to the conservation of energy and the spreading of light rays over a larger surface area. Modern microscopes, known as *compound microscopes* have many lenses in them (typically four) to optimize the functionality and enhance image stability. A slightly different variety of microscope, the comparison microscope, looks at side-by-side images to produce a stereoscopic binocular view that appears three dimensional when used by humans.

The first telescopes, called *refracting telescopes* were also developed with a single objective and eyepiece lens. In contrast to the microscope, the objective lens of the telescope was designed with a large focal length to avoid optical aberrations. The objective focuses an image of a distant object at its focal point which is adjusted to be at the focal point of an eyepiece of a much smaller focal length. The main goal of a telescope is not necessarily magnification, but rather collection of light which is determined by the physical size of the objective lens. Thus, telescopes are normally indicated by the diameters of their objectives rather than by the magnification which can be changed by switching eyepieces. Because the magnification of a telescope is equal to the focal length of the objective divided by the focal length of the eyepiece, smaller focal-length eyepieces cause greater magnification.

Since crafting large lenses is much more difficult than crafting large mirrors, most modern telescopes are *reflecting telescopes*, that is, telescopes that use a primary mirror rather than an objective lens. The same general optical considerations apply to reflecting telescopes that applied to refracting telescopes, namely, the larger the primary mirror, the more light collected, and the magnification is still equal to the focal length of the primary mirror divided by the focal length of the eyepiece. Professional telescopes generally do not have eyepieces and instead place an instrument (often a charge-coupled device) at the focal point instead.

Photography

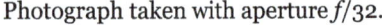

Photograph taken with aperture $f/32$.   Photograph taken with aperture $f/5$.

The optics of photography involves both lenses and the medium in which the electromagnetic radiation is recorded, whether it be a plate, film, or charge-coupled device. Photographers must consider the reciprocity of the camera and the shot which is summarized by the relation:

$$\text{Exposure} \propto \text{ApertureArea} \times \text{ExposureTime} \times \text{SceneLuminance}$$

In other words, the smaller the aperture (giving greater depth of focus), the less light coming in, so the length of time has to be increased (leading to possible blurriness if motion occurs). An example of the use of the law of reciprocity is the Sunny 16 rule which gives a rough estimate for the settings needed to estimate the proper exposure in daylight.

A camera's aperture is measured by a unitless number called the f-number or f-stop, $f/\#$, often notated as N, and given by,

$$f/\# = N = \frac{f}{D}$$

where f is the focal length, and D is the diameter of the entrance pupil. By convention, "$f/\#$" is treated as a single symbol, and specific values of $f/\#$ are written by replacing the number sign with the value. The two ways to increase the f-stop are to either decrease the diameter of the entrance pupil or change to a longer focal length (in the case of a zoom lens, this can be done by simply adjusting the lens). Higher f-numbers also have a larger depth of field due to the lens approaching the limit of a pinhole camera which is able to focus all images perfectly, regardless of distance, but requires very long exposure times.

The field of view that the lens will provide changes with the focal length of the lens. There are three basic classifications based on the relationship to the diagonal size of the film or sensor size of the camera to the focal length of the lens:

- Normal lens: Angle of view of about 50° (called *normal* because this angle considered roughly equivalent to human vision) and a focal length approximately equal to the diagonal of the film or sensor.

- Wide-angle lens: Angle of view wider than 60° and focal length shorter than a normal lens.

- Long focus lens: Angle of view narrower than a normal lens. This is any lens with a focal length longer than the diagonal measure of the film or sensor. The most common type of long focus lens is the telephoto lens, a design that uses a special *telephoto group* to be physically shorter than its focal length.

Modern zoom lenses may have some or all of these attributes.

The absolute value for the exposure time required depends on how sensitive to light the medium being used is (measured by the film speed, or, for digital media, by the quantum efficiency). Early photography used media that had very low light sensitivity, and so exposure times had to be long even for very bright shots. As technology has improved, so has the sensitivity through film cameras and digital cameras.

Other results from physical and geometrical optics apply to camera optics. For example, the maximum resolution capability of a particular camera set-up is determined by the diffraction limit associated with the pupil size and given, roughly, by the Rayleigh criterion.

Atmospheric Optics

The unique optical properties of the atmosphere cause a wide range of spectacular optical phenomena. The blue colour of the sky is a direct result of Rayleigh scattering which redirects higher frequency (blue) sunlight back into the field of view of the observer. Because blue light is scattered more easily than red light, the sun takes on a reddish hue when it is observed through a thick atmosphere, as during a sunrise or sunset. Additional particulate matter in the sky can scatter different colours at different angles creating colourful glowing skies at dusk and dawn. Scattering off of ice crystals and other particles in the atmosphere are responsible for halos, afterglows, coronas, rays of sunlight, and sun dogs. The variation in these kinds of phenomena is due to different particle sizes and geometries.

Mirages are optical phenomena in which light rays are bent due to thermal variations in the refraction index of air, producing displaced or heavily distorted images of distant objects. Other dramatic optical phenomena associated with this include the Novaya Zemlya effect where the sun appears to rise earlier than predicted with a distorted shape. A spectacular form of refraction occurs with a temperature inversion called the Fata Morgana where objects on the horizon or even beyond the horizon, such as islands, cliffs, ships or icebergs, appear elongated and elevated, like "fairy tale castles".

A colourful sky is often due to scattering of light off particulates and pollution, as in this photograph of a sunset during the October 2007 California wildfires.

Rainbows are the result of a combination of internal reflection and dispersive refraction of light in raindrops. A single reflection off the backs of an array of raindrops produces a rainbow with an angular size on the sky that ranges from 40° to 42° with red on the outside. Double rainbows are produced by two internal reflections with angular size of 50.5° to 54° with violet on the outside. Because rainbows are seen with the sun 180° away from the centre of the rainbow, rainbows are more prominent the closer the sun is to the horizon.

ACOUSTICS

Acoustics is the branch of physics concerned with the production, control, transmission, reception, and effects of sound. The term is derived from the Greek akoustos, meaning "heard."

Beginning with its origins in the study of mechanical vibrations and the radiation of these vibrations through mechanical waves, acoustics has had important applications in almost every area of life. It has been fundamental to many developments in the arts—some of which, especially in the area of musical scales and instruments, took place after long experimentation by artists and were only much later explained as theory by scientists. For example, much of what is now known about architectural acoustics was actually learned by trial and error over centuries of experience and was only recently formalized into a science.

Other applications of acoustic technology are in the study of geologic, atmospheric, and underwater phenomena. Psychoacoustics, the study of the physical effects of sound on biological systems, has been of interest since Pythagoras first heard the sounds of vibrating strings and of hammers hitting anvils in the 6th century BC, but the application of modern ultrasonic technology has only recently provided some of the most exciting developments in medicine. Even today, research continues into many aspects of the fundamental physical processes involved in waves and sound and into possible applications of these processes in modern life.

Early Experimentation

The origin of the science of acoustics is generally attributed to the Greek philosopher Pythagoras, whose experiments on the properties of vibrating strings that produce pleasing musical intervals were of such merit that they led to a tuning system that bears his name. Aristotle correctly suggested that a sound wave propagates in air through motion of the air—a hypothesis based more on philosophy than on experimental physics; however, he also incorrectly suggested that high frequencies propagate faster than low frequencies—an error that persisted for many centuries. Vitruvius, a Roman architectural engineer of the 1st century BC, determined the correct mechanism for the transmission of sound waves, and he contributed substantially to the acoustic design of theatres. In the 6th century AD, the Roman philosopher Boethius documented several ideas relating science to music, including a suggestion that the human perception of pitch is related to the physical property of frequency.

The modern study of waves and acoustics is said to have originated with Galileo Galilei, who elevated to the level of science the study of vibrations and the correlation between pitch and frequency of the sound source. His interest in sound was inspired in part by his father, who was a mathematician, musician, and composer of some repute. Following Galileo's foundation work, progress in acoustics came relatively rapidly. The French mathematician Marin Mersenne studied the vibration of stretched strings; the results of these studies were summarized in the three Mersenne's laws. Mersenne's Harmonicorum Libri provided the basis for modern musical acoustics. Later in the century Robert Hooke, an English physicist, first produced a sound wave of known frequency, using a rotating cog wheel as a measuring device. Further developed in the 19th century by the French physicist Félix Savart, and now commonly called Savart's disk, this device is often used today for demonstrations during physics lectures. In the late 17th and early 18th centuries, detailed studies of the relationship between frequency and pitch and of waves in stretched strings were carried out by the French physicist Joseph Sauveur, who provided a legacy of acoustic terms used to this day and first suggested the name acoustics for the study of sound.

One of the most interesting controversies in the history of acoustics involves the famous and often misinterpreted "bell-in-vacuum" experiment, which has become a staple of contemporary physics lecture demonstrations. In this experiment the air is pumped out of a jar in which a ringing bell is located; as air is pumped out, the sound of the bell diminishes until it becomes inaudible. As late as the 17th century many philosophers and scientists believed that sound propagated via invisible particles originating at the source of the sound and moving through space to affect the ear of the observer. The concept of sound as a wave directly challenged this view, but it was not established experimentally until the first bell-in-vacuum experiment was performed by Athanasius Kircher, a German scholar, who described it in his book Musurgia Universalis. Even after pumping the air out of the jar, Kircher could still hear the bell, so he concluded incorrectly that air was not required to transmit sound. In fact, Kircher's jar was not entirely free of air, probably because of inadequacy in his vacuum pump. By 1660 the Anglo-Irish scientist Robert Boyle had improved vacuum technology to the point where he could observe sound intensity decreasing virtually to zero as air was pumped out. Boyle then came to the correct conclusion that a medium such as air is required for transmission of sound waves. Although this conclusion is correct, as an explanation for the results of the bell-in-vacuum experiment it is misleading. Even with the mechanical pumps of today, the amount of air remaining in a vacuum jar is more than sufficient to transmit a sound wave. The real reason for a decrease in sound level upon pumping air out of the jar is that the bell is unable to transmit the sound vibrations efficiently to the less dense air remaining, and that air is likewise unable to transmit the sound efficiently to the glass jar. Thus, the real problem is one of an impedance mismatch between the air and the denser solid materials—and not the lack of a medium such as air, as is generally presented in textbooks. Nevertheless, despite the confusion regarding this experiment, it did aid in establishing sound as a wave rather than as particles.

Measuring the Speed of Sound

Once it was recognized that sound is in fact a wave, measurement of the speed of sound became a serious goal. In the 17th century, the French scientist and philosopher Pierre Gassendi made the earliest known attempt at measuring the speed of sound in air. Assuming correctly that the speed of light is effectively infinite compared with the speed of sound, Gassendi measured the time difference between spotting the flash of a gun and hearing its report over a long distance on a still day. Although the value he obtained was too high—about 478.4 metres per second (1,569.6 feet per second)—he correctly concluded that the speed of sound is independent of frequency. In the 1650s, Italian physicists Giovanni Alfonso Borelli and Vincenzo Viviani obtained the much better value of 350 metres per second using the same technique. Their compatriot G.L. Bianconi demonstrated in 1740 that the speed of sound in air increases with temperature. The earliest precise experimental value for the speed of sound, obtained at the Academy of Sciences in Paris in 1738, was 332 metres per second—incredibly close to the presently accepted value, considering the rudimentary nature of the measuring tools of the day. A more recent value for the speed of sound, 331.45 metres per second (1,087.4 feet per second), was obtained in 1942; it was amended in 1986 to 331.29 metres per second at 0° C (1,086.9 feet per second at 32° F).

The speed of sound in water was first measured by Daniel Colladon, a Swiss physicist, in 1826. Strangely enough, his primary interest was not in measuring the speed of sound in water but in calculating water's compressibility—a theoretical relationship between the speed of sound in a material and the material's compressibility having been established previously. Colladon came up with a speed of 1,435 metres per second at 8° C; the presently accepted value interpolated at that temperature is about 1,439 metres per second.

Two approaches were employed to determine the velocity of sound in solids. In 1808 Jean-Baptiste Biot, a French physicist, conducted direct measurements of the speed of sound in 1,000 metres of iron pipe by comparing it with the speed of sound in air. A better measurement had earlier been carried out by a German, Ernst Florenz Friedrich Chladni, using analysis of the nodal pattern in standing-wave vibrations in long rods.

Modern Advances

Simultaneous with these early studies in acoustics, theoreticians were developing the mathematical theory of waves required for the development of modern physics, including acoustics. In the early 18th century, the English mathematician Brook Taylor developed a mathematical theory of vibrating strings that agreed with previous experimental observations, but he was not able to deal with vibrating systems in general without the proper mathematical base. This was provided by Isaac Newton of England and Gottfried Wilhelm Leibniz of Germany, who, in pursuing other interests, independently developed the theory of calculus, which in turn allowed the derivation of the general wave equation by the French mathematician and scientist Jean Le Rond d'Alembert in the 1740s. The Swiss mathematicians Daniel Bernoulli and Leonhard Euler, as well as the Italian-French mathematician Joseph-Louis Lagrange, further applied the new equations of calculus to waves in strings and in the air. In the 19th century, Siméon-Denis Poisson of France extended these developments to stretched membranes, and the German mathematician Rudolf Friedrich Alfred Clebsch completed Poisson's earlier studies. A German experimental physicist, August Kundt, developed a number of important techniques for investigating properties of sound waves.

One of the most important developments in the 19th century involved the theory of vibrating plates. In addition to his work on the speed of sound in metals, Chladni had earlier introduced a technique of observing standing-wave patterns on vibrating plates by sprinkling sand onto the plates—a demonstration commonly used today. Perhaps the most significant step in the theoretical explanation of these vibrations was provided in 1816 by the French mathematician Sophie Germain, whose explanation was of such elegance and sophistication that errors in her treatment of the problem were not recognized until some 35 years later, by the German physicist Gustav Robert Kirchhoff.

The analysis of a complex periodic wave into its spectral components was theoretically established early in the 19th century by Jean-Baptiste-Joseph Fourier of France and is now commonly referred to as the Fourier theorem. The German physicist Georg Simon Ohm first suggested that the ear is sensitive to these spectral components; his idea that the ear is sensitive to the amplitudes but not the phases of the harmonics of a complex tone is known as Ohm's law of hearing (distinguishing it from the more famous Ohm's law of electrical resistance).

Hermann von Helmholtz made substantial contributions to understanding the mechanisms of hearing and to the psychophysics of sound and music. His book On the Sensations of Tone As a Physiological Basis for the Theory of Music is one of the classics of acoustics. In addition, he constructed a set of resonators, covering much of the audio spectrum, which were used in the spectral analysis of musical tones. The Prussian physicist Karl Rudolph Koenig, an extremely clever and creative experimenter, designed many of the instruments used for research in hearing and music, including a frequency standard and the manometric flame. The flame-tube device, used to render standing sound waves "visible," is still one of the most fascinating of physics classroom demonstrations. The English physical scientist John William Strutt, 3rd Baron Rayleigh, carried out an enormous variety of acoustic research; much of it was included in his two-volume treatise, The Theory of Sound, publication of which in 1877–78 is now thought to mark the beginning of modern acoustics. Much of Rayleigh's work is still directly quoted in contemporary physics textbooks.

The study of ultrasonics was initiated by the American scientist John LeConte, who in the 1850s developed a technique for observing the existence of ultrasonic waves with a gas flame. This technique was later used by the British physicist John Tyndall for the detailed study of the properties of sound waves. The piezoelectric effect, a primary means of producing and sensing ultrasonic waves, was discovered by the French physical chemist Pierre Curie and his brother Jacques in 1880. Applications of ultrasonics, however, were not possible until the development in the early 20th century of the electronic oscillator and amplifier, which were used to drive the piezoelectric element.

Among 20th-century innovators were the American physicist Wallace Sabine, considered to be the originator of modern architectural acoustics, and the Hungarian-born American physicist Georg von Békésy, who carried out experimentation on the ear and hearing and validated the commonly accepted place theory of hearing first suggested by Helmholtz. Békésy's book Experiments in Hearing, published in 1960, is the magnum opus of the modern theory of the ear.

Amplifying, Recording and Reproducing

The earliest known attempt to amplify a sound wave was made by Athanasius Kircher, of "bell-in-vacuum" fame; Kircher designed a parabolic horn that could be used either as a hearing aid or as a voice amplifier. The amplification of body sounds became an important goal, and the first stethoscope was invented by a French physician, René Laënnec, in the early 19th century.

Attempts to record and reproduce sound waves originated with the invention in 1857 of a mechanical sound-recording device called the phonautograph by Édouard-Léon Scott de Martinville. The first device that could actually record and play back sounds was developed by the American inventor Thomas Alva Edison in 1877. Edison's phonograph employed grooves of varying depth in a cylindrical sheet of foil, but a spiral groove on a flat rotating disk was introduced a decade later by the German-born American inventor Emil Berliner in an invention he called the gramophone. Much significant progress in recording and reproduction techniques was made during the first half of the 20th century, with the development of high-quality electromechanical transducers and linear electronic circuits. The most important improvement on the standard phonograph record in the second half of the century was the compact disc, which employed digital techniques developed in mid-century that substantially reduced noise and increased the fidelity and durability of the recording.

Architectural Acoustics

Reverberation Time

Although architectural acoustics has been an integral part of the design of structures for at least 2,000 years, the subject was only placed on a firm scientific basis at the beginning of the 20th century by Wallace Sabine. Sabine pointed out that the most important quantity in determining the acoustic suitability of a room for a particular use is its reverberation time, and he provided a scientific basis by which the reverberation time can be determined or predicted.

When a source creates a sound wave in a room or auditorium, observers hear not only the sound wave propagating directly from the source but also the myriad reflections from the walls, floor, and ceiling. These latter form the reflected wave, or reverberant sound. After the source ceases, the reverberant sound can be heard for some time as it grows softer. The time required, after the sound source ceases, for the absolute intensity to drop by a factor of 106—or, equivalently, the time for the intensity level to drop by 60 decibels—is defined as the reverberation time (RT, sometimes referred to as RT60). Sabine recognized that the reverberation time of an auditorium is related to the volume of the auditorium and to the ability of the walls, ceiling, floor, and contents of the room to absorb sound. Using these assumptions, he set forth the empirical relationship through which the reverberation time could be determined: RT = 0.05V/A, where RT is the reverberation time in seconds, V is the volume of the room in cubic feet, and A is the total sound absorption of the room, measured by the unit sabin. The sabin is the absorption equivalent to one square foot of perfectly absorbing surface—for example, a one-square-foot hole in a wall or five square feet of surface that absorbs 20 percent of the sound striking it.

Both the design and the analysis of room acoustics begin with this equation. Using the equation and the absorption coefficients of the materials from which the walls are to be constructed, an approximation can be obtained for the way in which the room will function acoustically. Absorbers and reflectors, or some combination of the two, can then be used to modify the reverberation time and its frequency dependence, thereby achieving the most desirable characteristics for specific uses. Representative absorption coefficients—showing the fraction of the wave, as a function of frequency, that is absorbed when a sound hits various materials—are given in the table. The absorption from all the surfaces in the room are added together to obtain the total absorption (A).

Absorption coefficients of common materials at several frequencies.						
Material	Frequency (hertz)					
	125	250	500	1,000	2,000	4,000
Concrete	0.01	0.01	0.02	0.02	0.02	0.03
Plasterboard	0.20	0.15	0.10	0.08	0.04	0.02
Acoustic board	0.25	0.45	0.80	0.90	0.90	0.90
Curtains	0.05	0.12	0.25	0.35	0.40	0.45

While there is no exact value of reverberation time that can be called ideal, there is a range of values deemed to be appropriate for each application. These vary with the size of the room, but the averages can be calculated and indicated by lines on a graph. The need for clarity in understanding speech dictates that rooms used for talking must have a reasonably short reverberation time. On the other hand, the full sound desirable in the performance of music of the Romantic era, such as Wagner operas or Mahler symphonies, requires a long reverberation time. Obtaining a clarity suitable for the light, rapid passages of Bach or Mozart requires an intermediate value of reverberation time. For playing back recordings on an audio system, the reverberation time should be short, so as not to create confusion with the reverberation time of the music in the hall where it was recorded.

Acoustic Criteria

Many of the acoustic characteristics of rooms and auditoriums can be directly attributed to specific physically measurable properties. Because the music critic or performing artist uses a different vocabulary to describe these characteristics than does the physicist, it is helpful to survey some of the more important features of acoustics and correlate the two sets of descriptions.

"Liveness" refers directly to reverberation time. A live room has a long reverberation time and a dead room a short reverberation time. "Intimacy" refers to the feeling that listeners have of being physically close to the performing group. A room is generally judged intimate when the first reverberant sound reaches the listener within about 20 milliseconds of the direct sound. This condition is met easily in a small room, but it can also be achieved in large halls by the use of orchestral shells that partially enclose the performers. Another example is a canopy placed above a speaker in a large room such as a cathedral: this leads to both a strong and a quick first reverberation and thus to a sense of intimacy with the person speaking.

The amplitude of the reverberant sound relative to the direct sound is referred to as fullness. Clarity, the opposite of fullness, is achieved by reducing the amplitude of the reverberant sound. Fullness generally implies a long reverberation time, while clarity implies a shorter reverberation time. A fuller sound is generally required of Romantic music or performances by larger groups, while more clarity would be desirable in the performance of rapid passages from Bach or Mozart or in speech.

"Warmth" and "brilliance" refer to the reverberation time at low frequencies relative to that at higher frequencies. Above about 500 hertz, the reverberation time should be the same for all frequencies. But at low frequencies an increase in the reverberation time creates a warm sound, while, if the reverberation time increased less at low frequencies, the room would be characterized as more brilliant.

"Texture" refers to the time interval between the arrival of the direct sound and the arrival of the first few reverberations. To obtain good texture, it is necessary that the first five reflections arrive at the observer within about 60 milliseconds of the direct sound. An important corollary to this requirement is that the intensity of the reverberations should decrease monotonically; there should be no unusually large late reflections.

"Blend" refers to the mixing of sounds from all the performers and their uniform distribution to the listeners. To achieve proper blend it is often necessary to place a collection of reflectors on the stage that distribute the sound randomly to all points in the audience.

Although the above features of auditorium acoustics apply to listeners, the idea of ensemble applies primarily to performers. In order to perform coherently, members of the ensemble must be able to hear one another. Reverberant sound cannot be heard by the members of an orchestra, for example, if the stage is too wide, has too high a ceiling, or has too much sound absorption on its sides.

Acoustic Problems

Certain acoustic problems often result from improper design or from construction limitations. If large echoes are to be avoided, focusing of the sound wave must be avoided. Smooth, curved reflecting surfaces such as domes and curved walls act as focusing elements, creating large echoes and leading to bad texture. Improper blend results if sound from one part of the ensemble is focused to one section of the audience. In addition, parallel walls in an auditorium reflect sound back and forth, creating a rapid, repetitive pulsing of sound known as flutter echo and even leading to destructive interference of the sound wave. Resonances at certain frequencies should also be avoided by use of oblique walls.

Acoustic shadows, regions in which some frequency regions of sound are attenuated, can be caused by diffraction effects as the sound wave passes around large pillars and corners or underneath a low balcony. Large reflectors called clouds, suspended over the performers, can be of such a size as to reflect certain frequency regions while allowing others to pass, thus affecting the mixture of the sound.

External noise can be a serious problem for halls in urban areas or near airports or highways. One technique often used for avoiding external noise is to construct the auditorium as a smaller room within a larger room. Noise from air blowers or other mechanical vibrations can be reduced using techniques involving impedance and by isolating air handlers.

Good acoustic design must take account of all these possible problems while emphasizing the desired acoustic features. One of the problems in a large auditorium involves simply delivering an adequate amount of sound to the rear of the hall. The intensity of a spherical sound wave decreases in intensity at a rate of six decibels for each factor of two increase in distance from the source, as shown above. If the auditorium is flat, a hemispherical wave will result. Absorption of the diffracted wave by the floor or audience near the bottom of the hemisphere will result in even greater absorption, so that the resulting intensity level will fall off at twice the theoretical rate, at about 12 decibels for each factor of two in distance. Because of this absorption, the floors of an auditorium are generally sloped upward toward the rear.

THERMODYNAMICS

Thermodynamics is the branch of physics that deals with heat and temperature, and their relation to energy, work, radiation, and properties of matter. The behavior of these quantities is governed by the four laws of thermodynamics which convey a quantitative description using measurable macroscopic physical quantities, but may be explained in terms of microscopic constituents by statistical mechanics. Thermodynamics applies to a wide variety of topics in science and engineering, especially physical chemistry, chemical engineering and mechanical engineering, but also in fields as complex as meteorology.

Historically, thermodynamics developed out of a desire to increase the efficiency of early steam engines, particularly through the work of French physicist Nicolas Léonard Sadi Carnot who believed that engine efficiency was the key that could help France win the Napoleonic Wars. Scots-Irish physicist Lord Kelvin was the first to formulate a concise definition of thermodynamics in 1854 which stated, "Thermo-dynamics is the subject of the relation of heat to forces acting between contiguous parts of bodies, and the relation of heat to electrical agency."

The initial application of thermodynamics to mechanical heat engines was extended early on to the study of chemical compounds and chemical reactions. Chemical thermodynamics studies the nature of the role of entropy in the process of chemical reactions and has provided the bulk of expansion and knowledge of the field. Other formulations of thermodynamics emerged. Statistical thermodynamics, or statistical mechanics, concerns itself with statistical predictions of the collective motion of particles from their microscopic behavior. In 1909, Constantin Carathéodory presented a purely mathematical approach in an axiomatic formulation, a description often referred to as *geometrical thermodynamics*.

A description of any thermodynamic system employs the four laws of thermodynamics that form an axiomatic basis. The first law specifies that energy can be exchanged between physical systems as heat and work. The second law defines the existence of a quantity called entropy, that describes the direction, thermodynamically, that a system can evolve and quantifies the state of order of a system and that can be used to quantify the useful work that can be extracted from the system.

In thermodynamics, interactions between large ensembles of objects are studied and categorized. Central to this are the concepts of the thermodynamic *system* and its *surroundings*. A system is composed of particles, whose average motions define its properties, and those properties are in turn related to one another through equations of state. Properties can be combined to express internal energy and thermodynamic potentials, which are useful for determining conditions for equilibrium and spontaneous processes.

With these tools, thermodynamics can be used to describe how systems respond to changes in their environment. This can be applied to a wide variety of topics in science and engineering, such as engines, phase transitions, chemical reactions, transport phenomena, and even black holes. The results of thermodynamics are essential for other fields of physics and for chemistry, chemical engineering, corrosion engineering, aerospace engineering, mechanical engineering, cell biology, biomedical engineering, materials science, and economics, to name a few.

This topic is focused mainly on classical thermodynamics which primarily studies systems in thermodynamic equilibrium. Non-equilibrium thermodynamics is often treated as an extension of the classical treatment, but statistical mechanics has brought many advances to that field.

Branches of Thermodynamics

The study of thermodynamical systems has developed into several related branches, each using a different fundamental model as a theoretical or experimental basis, or applying the principles to varying types of systems.

Classical Thermodynamics

Classical thermodynamics is the description of the states of thermodynamic systems at near-equilibrium, that uses macroscopic, measurable properties. It is used to model exchanges of energy, work and heat based on the laws of thermodynamics. The qualifier *classical* reflects the fact that it represents the first level of understanding of the subject as it developed in the 19th century and describes the changes of a system in terms of macroscopic empirical (large scale, and measurable) parameters. A microscopic interpretation of these concepts was later provided by the development of *statistical mechanics*.

Statistical Mechanics

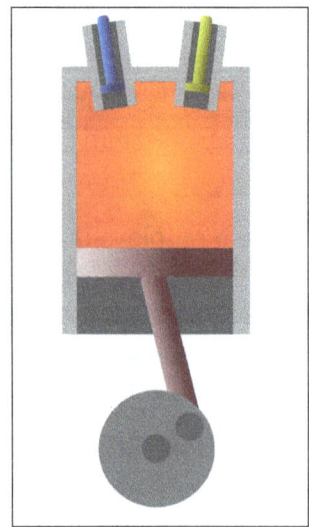

Statistical mechanics, also called statistical thermodynamics, emerged with the development of atomic and molecular theories in the late 19th century and early 20th century, and supplemented classical thermodynamics with an interpretation of the microscopic interactions between individual particles or quantum-mechanical states. This field relates the microscopic properties of individual atoms and molecules to the macroscopic, bulk properties of materials that can be observed on the human scale, thereby explaining classical thermodynamics as a natural result of statistics, classical mechanics, and quantum theory at the microscopic level.

Chemical Thermodynamics

Chemical thermodynamics is the study of the interrelation of energy with chemical reactions or with a physical change of state within the confines of the laws of thermodynamics.

Equilibrium Thermodynamics

Equilibrium thermodynamics is the systematic study of transfers of matter and energy in systems as they pass from one state of thermodynamic equilibrium to another. The term 'thermodynamic equilibrium' indicates a state of balance. In an equilibrium state there are no unbalanced potentials, or driving forces, between macroscopically distinct parts of the system. A central aim in equilibrium thermodynamics is: given a system in a well-defined initial equilibrium state, and given its surroundings, and given its constitutive walls, to calculate what will be the final equilibrium state of the system after a specified thermodynamic operation has changed its walls or surroundings.

Non-equilibrium thermodynamics is a branch of thermodynamics that deals with systems that are not in thermodynamic equilibrium. Most systems found in nature are not in thermodynamic equilibrium because they are not in stationary states, and are continuously and discontinuously subject to flux of matter and energy to and from other systems. The thermodynamic study of non-equilibrium systems requires more general concepts than are dealt with by equilibrium thermodynamics. Many natural systems still today remain beyond the scope of currently known macroscopic thermodynamic methods.

Laws of Thermodynamics

Thermodynamics is principally based on a set of four laws which are universally valid when applied to systems that fall within the constraints implied by each. In the various theoretical descriptions of thermodynamics these laws may be expressed in seemingly differing forms, but the most prominent formulations are the following:

- Zeroth law of thermodynamics: If two systems are each in thermal equilibrium with a third, they are also in thermal equilibrium with each other.

This statement implies that thermal equilibrium is an equivalence relation on the set of thermodynamic systems under consideration. Systems are said to be in equilibrium if the small, random exchanges between them (e.g. Brownian motion) do not lead to a net change in energy. This law is tacitly assumed in every measurement of temperature. Thus, if one seeks to decide if two bodies are at the same temperature, it is not necessary to bring them into contact and measure any changes of their observable properties in time. The law provides an empirical definition of temperature and justification for the construction of practical thermometers.

The zeroth law was not initially named as a law of thermodynamics, as its basis in thermodynamical equilibrium was implied in the other laws. The first, second, and third laws had been explicitly stated prior and found common acceptance in the physics community. Once the importance of the zeroth law for the definition of temperature was realized, it was impracticable to renumber the other laws, hence it was numbered the *zeroth law*.

- First law of thermodynamics: The internal energy of an isolated system is constant.

The first law of thermodynamics is an expression of the principle of conservation of energy. It states that energy can be transformed (changed from one form to another), but cannot be created or destroyed.

The first law is usually formulated by saying that the change in the internal energy of a closed thermodynamic system is equal to the difference between the heat supplied to the system and the amount of work done by the system on its surroundings. It is important to note that internal energy is a state of the system whereas heat and work modify the state of the system. In other words, a change of internal energy of a system may be achieved by any combination of heat and work added or removed from the system as long as those total to the change of internal energy. The manner by which a system achieves its internal energy is path independent.

- Second law of thermodynamics: Heat cannot spontaneously flow from a colder location to a hotter location.

The second law of thermodynamics is an expression of the universal principle of decay observable in nature. The second law is an observation of the fact that over time, differences in temperature, pressure, and chemical potential tend to even out in a physical system that is isolated from the outside world. Entropy is a measure of how much this process has progressed. The entropy of an isolated system which is not in equilibrium will tend to increase over time, approaching a maximum value at equilibrium. However, principles guiding systems that are far from equilibrium are still debatable. One of such principles is the maximum entropy production principle. It states that non-equilibrium systems behave such a way as to maximize its entropy production.

In classical thermodynamics, the second law is a basic postulate applicable to any system involving heat energy transfer; in statistical thermodynamics, the second law is a consequence of the assumed randomness of molecular chaos. There are many versions of the second law, but they all have the same effect, which is to explain the phenomenon of irreversibility in nature.

- Third law of thermodynamics: As a system approaches absolute zero, all processes cease and the entropy of the system approaches a minimum value.

The third law of thermodynamics is a statistical law of nature regarding entropy and the impossibility of reaching absolute zero of temperature. This law provides an absolute reference point for the determination of entropy. The entropy determined relative to this point is the absolute entropy. Alternate definitions are, "the entropy of all systems and of all states of a system is smallest at absolute zero," or equivalently "it is impossible to reach the absolute zero of temperature by any finite number of processes".

Absolute zero, at which all activity would stop if it were possible to happen, is −273.15 °C (degrees Celsius), or −459.67 °F (degrees Fahrenheit), or 0 K (kelvin), or 0° R (degrees Rankine).

ASTROPHYSICS

Astrophysics is a branch of space science that applies the laws of physics and chemistry to explain the birth, life and death of stars, planets, galaxies, nebulae and other objects in the universe. It has two sibling sciences, astronomy and cosmology, and the lines between them blur.

In the most rigid sense:

- Astronomy measures positions, luminosities, motions and other characteristics.

- Astrophysics creates physical theories of small to medium-size structures in the universe.

- Cosmology does this for the largest structures, and the universe as a whole.

In practice, the three professions form a tight-knit family. Ask for the position of a nebula or what kind of light it emits, and the astronomer might answer first. Ask what the nebula is made of and how it formed and the astrophysicist will pipe up. Ask how the data fit with the formation of the universe, and the cosmologist would probably jump in. But watch out — for any of these questions, two or three may start talking at once.

Goals of Astrophysics

Astrophysicists seek to understand the universe and our place in it. At NASA, the goals of astrophysics are "to discover how the universe works, explore how it began and evolved, and search for life on planets around other stars," according NASA's website.

NASA states that those goals produce three broad questions:

- How does the universe work?

- How did we get here?

- Are we alone?

It began with Newton

While astronomy is one of the oldest sciences, theoretical astrophysics began with Isaac Newton. Prior to Newton, astronomers described the motions of heavenly bodies using complex mathematical models without a physical basis. Newton showed that a single theory simultaneously explains the orbits of moons and planets in space and the trajectory of a cannonball on Earth. This added to the body of evidence for the (then) startling conclusion that the heavens and Earth are subject to the same physical laws.

Perhaps what most completely separated Newton's model from previous ones is that it is predictive as well as descriptive. Based on aberrations in the orbit of Uranus, astronomers predicted the position of a new planet, which was then observed and named Neptune. Being predictive as well as descriptive is the sign of a mature science, and astrophysics is in this category.

Milestones in Astrophysics

Because the only way we interact with distant objects is by observing the radiation they emit, much of astrophysics has to do with deducing theories that explain the mechanisms that produce this radiation, and provide ideas for how to extract the most information from it. The first ideas about the nature of stars emerged in the mid-19th century from the blossoming science of spectral analysis, which means observing the specific frequencies of light that particular substances absorb and emit when heated. Spectral analysis remains essential to the triumvirate of space sciences, both guiding and testing new theories.

Early spectroscopy provided the first evidence that stars contain substances also present on Earth. Spectroscopy revealed that some nebulae are purely gaseous, while some contain stars. This later helped cement the idea that some nebulae were not nebulae at all — they were other galaxies.

In the early 1920s, Cecilia Payne discovered, using spectroscopy, that stars are predominantly hydrogen (at least until their old age). The spectra of stars also allowed astrophysicists to determine the speed at which they move toward or away from Earth. Just like the sound a vehicle emits is different moving toward us or away from us, because of the Doppler shift, the spectra of stars will change in the same way. In the 1930s, by combining the Doppler shift and Einstein's theory of general relativity, Edwin Hubble provided solid evidence that the universe is expanding. This is also predicted by Einstein's theory, and together form the basis of the Big Bang Theory.

Also in the mid-19th century, the physicists Lord Kelvin (William Thomson) and Gustav Von Helmholtz speculated that gravitational collapse could power the sun, but eventually realized that energy produced this way would only last 100,000 years. Fifty years later, Einstein's famous $E=mc^2$ equation gave astrophysicists the first clue to what the true source of energy might be (although it turns out that gravitational collapse does play an important role). As nuclear physics, quantum mechanics and particle physics grew in the first half of the 20th century, it became possible to formulate theories for how nuclear fusion could power stars. These theories describe how stars form, live and die, and successfully explain the observed distribution of types of stars, their spectra, luminosities, ages and other features.

Astrophysics is the physics of stars and other distant bodies in the universe, but it also hits close to home. According to the Big Bang Theory, the first stars were almost entirely hydrogen. The nuclear fusion process that energizes them smashes together hydrogen atoms to form the heavier element helium. In 1957, the husband-and-wife astronomer team of Geoffrey and Margaret Burbidge, along with physicists William Alfred Fowler and Fred Hoyle, showed how, as stars age, they produce heavier and heavier elements, which they pass on to later generations of stars in ever-greater quantities. It is only in the final stages of the lives of more recent stars that the elements making up the Earth, such as iron (32.1 percent), oxygen (30.1 percent), silicon (15.1 percent), are produced. Another of these elements is carbon, which together with oxygen, make up the bulk of the mass of all living things, including us. Thus, astrophysics tells us that, while we are not all stars, we are all stardust.

References

- Meshik, A. P. (November 2005). "The Workings of an Ancient Nuclear Reactor". Scientific American. 293 (5): 82–91. Bibcode:2005sciam.293e..82M. Doi:10.1038/scientificamerican1105-82. Retrieved 2014-01-04

- Physics-science, science: britannica.com, Retrieved 14 February, 2019

- "World›s oldest telescope?". BBC News. July 1, 1999. Archived from the original on February 1, 2009. Retrieved Jan 3, 2010

- Atomic-physics, entry: newworldencyclopedia.org, Retrieved 15 March, 2019

- B. R. Martin (2006). Nuclear and Particle Physics. John Wiley & Sons, Ltd. ISBN 978-0-470-01999-3

- Acoustics, science: britannica.com, Retrieved 16 April, 2019

- Belkin, Andrey; et., al. (2015). "Self-Assembled Wiggling Nano-Structures and the Principle of Maximum Entropy Production". Sci. Rep. 5: 8323. Bibcode:2015natsr...5E8323B. Doi:10.1038/srep08323. PMC 4321171. PMID 25662746

- Astrophysics-26218: space.com, Retrieved 17 May, 2019

2

Force, Motion and Energy

Force is an interaction that changes the motion of the object. Motion is the action of changing the position of an object in a given interval of time. The quantitative property that is transferred to an object to perform work or to heat is referred as energy. The chapter closely examines key concepts of force, motion and energy to provide an extensive understanding of the subject.

FORCE

A 'Force' is a vector quantity that can be described as a push or pull on an object resulting from the object's interaction with another object. Whenever there is an interaction between two objects, the objects experience an equal and opposing force on each other. In other words both the objects 'exert force' on each other. Force only exists as a result of an interaction. If there is no interaction, the objects no longer experience the force. Force is measured in units called Newtons (N), named after the famous scientist Sir Issac Newton.

Interactive Nature of Force

For force to exist, there must be an interaction between at least two objects. This interaction between the objects may or may not be physical. Pulling a rope or hitting a ball with a bat come physical interaction whereas forces like the magnetic force, electrostatic force, etc come under non-physical force.

A force has both magnitude and direction. The magnitude of the force expresses the strength of a force. To completely determine a force both direction and magnitude must be specified; since the direction or magnitude of the applied force changes its effects.

Addition and Subtraction of Forces

'Net Force' is the total force exerted.

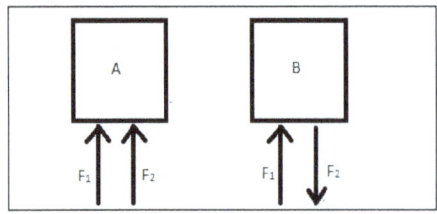

Therefore in the given diagram,

For the block A, the net force acting is F1+F2 . The two forces are added since the force is exerted in the same direction.

And, for the block B, the net force acting is F1−F2. The two forces are subtracted since the forces are applied in opposite directions.

Effects of Force

A force acting on an object causes the object to change its shape or size, to start moving, to stop moving, to accelerate or decelerate. When there's the interaction between two objects they exert a force on each other, these exerted forces are equal in size but opposite in direction. When an object has several forces acting on it, the effects of force is same as one force acting on the object in a certain direction and this overall force is called the 'resultant force'. The resultant force is essential to change the velocity of an object.

- If the resultant force is zero the forces on the object are balanced.

- If the resultant force acting on the object is 'zero' then: the object will remain stationary. In such a case the object will move at a steady speed in a straight line.

- If the resultant force acting on the object isn't zero then: the object will either accelerate or decelerate.

CONTACT FORCE

Friction

Friction is the force that resists the sliding or rolling of one solid object over another. Frictional forces, such as the traction needed to walk without slipping, may be beneficial, but they also present a great measure of opposition to motion. About 20 percent of the engine power of automobiles is consumed in overcoming frictional forces in the moving parts.

The major cause of friction between metals appears to be the forces of attraction, known as adhesion, between the contact regions of the surfaces, which are always microscopically irregular. Friction arises from shearing these "welded" junctions and from the action of the irregularities of the harder surface plowing across the softer surface.

Two simple experimental facts characterize the friction of sliding solids. First, the amount of friction is nearly independent of the area of contact. If a brick is pulled along a table, the frictional force is the same whether the brick is lying flat or standing on end. Second, friction is proportional to the load or weight that presses the surfaces together. If a pile of three bricks is pulled along a table, the friction is three times greater than if one brick is pulled. Thus, the ratio of friction F to load L is constant. This constant ratio is called the coefficient of friction and is usually symbolized by the Greek letter mu (μ). Mathematically, $\mu = F/L$. Because both friction and load are measured in units of force (such as pounds or newtons), the coefficient of friction is dimensionless. The value of the coefficient of friction for a case of one or more bricks sliding on a clean wooden table is about 0.5, which implies that a force equal to half the weight of the bricks is required just to overcome friction in keeping the bricks moving along at a constant speed. The frictional force itself is directed oppositely to the motion of the object. Because the friction thus far described arises between surfaces in relative motion, it is called kinetic friction.

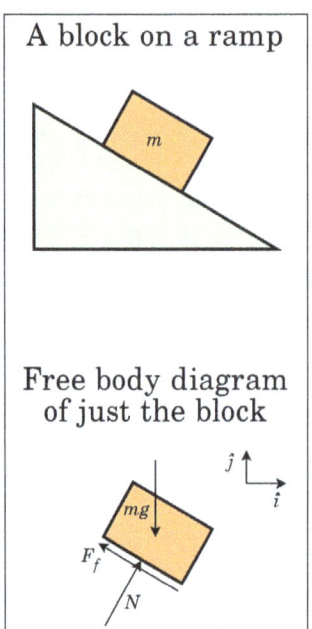

Block on a ramp and corresponding free body diagram of the block showing the contact force from the ramp onto the bottom of the block and separated into two components, a normal force N and a friction force f, along with the body force of gravity mg acting at the center of mass.

Static friction, in contrast, acts between surfaces at rest with respect to each other. The value of static friction varies between zero and the smallest force needed to start motion. This smallest force required to start motion, or to overcome static friction, is always greater than the force required to continue the motion, or to overcome kinetic friction.

Rolling friction occurs when a wheel, ball, or cylinder rolls freely over a surface, as in ball and roller bearings. The main source of friction in rolling appears to be dissipation of energy involved in deformation of the objects. If a hard ball is rolling on a level surface, the ball is somewhat flattened and the level surface somewhat indented in the regions in contact. The elastic deformation or compression produced at the leading section of the area in contact is a hindrance to motion that is not fully compensated as the substances spring back to normal shape at the trailing section. The internal losses in the two substances are similar to those that keep a ball from bouncing back to the level from which it is dropped. Coefficients of sliding friction are generally 100 to 1,000 times greater than coefficients of rolling friction for corresponding materials. This advantage was realized historically with the transition from sledge to wheel.

Tension

In physics, tension may be described as the pulling force transmitted axially by the means of a string, cable, chain, or similar one-dimensional continuous object, or by each end of a rod, truss member, or similar three-dimensional object; tension might also be described as the action-reaction pair of forces acting at each end of said elements. Tension could be the opposite of compression.

At the atomic level, when atoms or molecules are pulled apart from each other and gain potential energy with a restoring force still existing, the restoring force might create what is also called tension. Each end of a string or rod under such tension could pull on the object it is attached to, in order to restore the string/rod to its relaxed length.

In physics, tension, as a transmitted force, as an action-reaction pair of forces, or as a restoring force, may be a force and has the units of force measured in newtons (or sometimes pounds-force). The ends of a string or other object transmitting tension will exert forces on the objects to which the string or rod is connected, in the direction of the string at the point of attachment. These forces due to tension are also called "passive forces". There are two basic possibilities for systems of objects held by strings: either acceleration is zero and the system is therefore in equilibrium, or there is acceleration, and therefore a net force is present in the system.

Tension in One Dimension

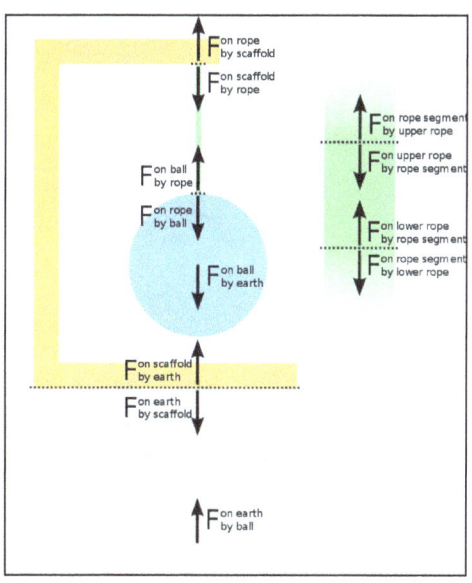

Tension in a string is a scalar quantity (i.e. non-negative). Zero tension is slack. A string or rope is often idealized as one dimension, having length but being massless with zero cross section. If there are no bends in the string, as occur with vibrations or pulleys, then tension is a constant along the string, equal to the magnitude of the forces applied by the ends of the string. By Newton's Third Law, these are the same forces exerted on the ends of the string by the objects to which the ends are attached. If the string curves around one or more pulleys, it will still have constant tension along its length in the idealized situation that the pulleys are massless and frictionless. A vibrating string vibrates with a set of frequencies that depend on the string's tension. These frequencies can be derived from Newton's laws of motion. Each microscopic segment of the string pulls on and is pulled upon by its neighboring segments, with a force equal to the tension at that position along the string. tension $= \tau(x)$ where x is the position along the string.

If the string has curvature, then the two pulls on a segment by its two neighbors will not add to zero, and there will be a net force on that segment of the string, causing an acceleration. This net force is a restoring force, and the motion of the string can include transverse waves that solve the equation central to Sturm-Liouville theory:

$$-\frac{d}{dx}\left[\tau(x)\frac{d\rho(x)}{dx}\right]+v(x)\rho(x)=\omega^2\sigma(x)\rho(x)$$

where $v(x)$ is the force constant per unit length [units force per area], ω^2 are the eigenvalues for resonances of transverse displacement $\rho(x)$ on the string., with solutions that include the various harmonics on a stringed instrument.

Tension in Three Dimensions

Tension is also used to describe the force exerted by the ends of a three-dimensional, continuous material such as a rod or truss member. Such a rod elongates under tension. The amount of elongation and the load that will cause failure both depend on the force per cross-sectional area rather than the force alone, so stress = axial force / cross sectional area is more useful for engineering purposes than tension. Stress is a 3x3 matrix called a tensor, and the $_{11}$ element of the stress tensor is tensile force per area, or compression force per area, denoted as a negative number for this element, if the rod is being compressed rather than elongated.

System in Equilibrium

A system is in equilibrium when the sum of all forces is zero.

$$\sum\vec{F}=0$$

For example, consider a system consisting of an object that is being lowered vertically by a string with tension, T, at a constant velocity. The system has a constant velocity and is therefore in equilibrium because the tension in the string, which is pulling up on the object, is equal to the weight force, mg ("m" is mass, "g" is the acceleration caused by the gravity of Earth), which is pulling down on the object.

$$\sum\vec{F}=\vec{T}+m\vec{g}=0$$

System under Net Force

A system has a net force when an unbalanced force is exerted on it, in other words the sum of all forces is not zero. Acceleration and net force always exist together.

$$\sum\vec{F}\neq0$$

For example, consider the same system as above but suppose the object is now being lowered with an increasing velocity downwards (positive acceleration) therefore there exists a net force somewhere in the system. In this case, negative acceleration would indicate that $|mg|>|T|$.

$$\sum \vec{F} = T - mg \neq 0$$

In another example, suppose that two bodies A and B having masses m_1 and m_2, respectively, are connected with each other by an inextensible string over a frictionless pulley. There are two forces acting on the body A: its weight ($w_1 = m_1 g$) pulling down, and the tension T in the string pulling up. Therefore, the net force F_1 on body A is $w_1 - T$, so $m_1 a = m_1 g - T$. In an extensible string, Hooke's law applies.

Strings in Modern Physics

String-like objects in relativistic theories, such as the strings used in some models of interactions between quarks, or those used in the modern string theory, also possess tension. These strings are analyzed in terms of their world sheet, and the energy is then typically proportional to the length of the string. As a result, the tension in such strings is independent of the amount of stretching.

NON-CONTACT FORCE

A non-contact force is a positive change in society which acts on an object without coming physically in contact with it. The most familiar example of a non-contact force is gravity, which confers weight. In contrast a contact force is a force applied to a body by another body that *is* in contact with it.

Examples:

All four known fundamental interactions are non-contact forces:

- Gravity, the force of attraction that exists among all bodies that have mass. The force exerted on each body by the other through weight is proportional to the mass of the first body times the mass of the second body divided by the square of the distance between them.

- Electromagnetism is the force that causes the interaction between electrically charged particles; the areas in which this happens are called electromagnetic fields. Examples of this force include: electricity, magnetism, radio waves, microwaves, infrared, visible light, X-rays and gamma rays. Electromagnetism mediates all chemical, biological, electrical and electronic processes.

- Strong nuclear force: Unlike gravity and electromagnetism, the strong nuclear force is a short distance force that takes place between fundamental particles within a nucleus. It is charge independent and acts equally between a proton and a proton, a neutron and a neutron, and a proton and a neutron. The strong nuclear force is the strongest force in nature; however, its range is small (acting only over distances of the order of 10^{-15} m). The strong nuclear force mediates both nuclear fission and fusion reactions.

- Weak nuclear force: The weak nuclear force mediates the β decay of a neutron, in which the neutron decays into a proton and in the process emits a β particle and an uncharged particle called a neutrino. As a result of mediating the β decay process, the weak nuclear force plays a key role in supernovas. Both the strong and weak forces form an important part of quantum mechanics.

MOTION

The continuous change in position of a body with respect to time and relative to the reference point or observer is called motion. Here the reference point can be considered as stationary or motionless objects surrounding the body.

Example :

Consider sun as reference point or observer the planets orbiting the sun are said to be in motion.

The position of a body is not changing with time and relative to the reference point then the body is said to be at rest or motionless. chairs, tables, Television are some examples of objects which are at rest. Rest and motion are relative terms for example consider a person in a moving bus he is at rest with respect to another person in the moving bus, but he is in motion with respect person out side the bus.

TRANSLATIONAL MOTION

Translational motion is the motion by which a body shifts from one point in space to another. One example of translational motion is the the motion of a bullet fired from a gun.

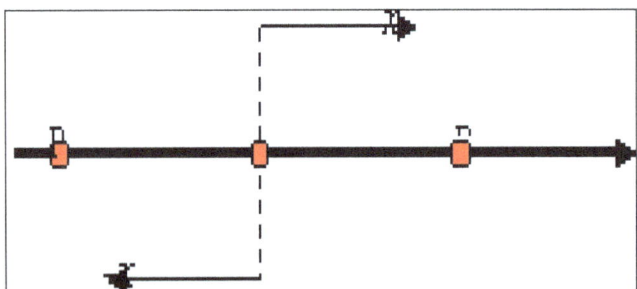

An object has a rectilinear motion when it moves along a straight line. At any time, t, the object occupies a position along the line as shown in the following figure. The distance x, with appropriate sign, define the position of the object. When the position of the object at particular time is known, the motion of the particle will be known, and generally is expressed in a form of an equation which relates distance x, to time t, for example x = 6t - 4, or a graph.

Motion in two or three dimensions is more complicated. In two dimensions, we need to specify two coordinates in order to fix the position of any object. The following figure shows a simple example of projectile motion: a ball rolling off a table. Let us define the horizontal direction as the x-axis and the vertical direction as the y-axis. Consider a ball initially rolling on off a flat table with an initial velocity of 10 m/s.

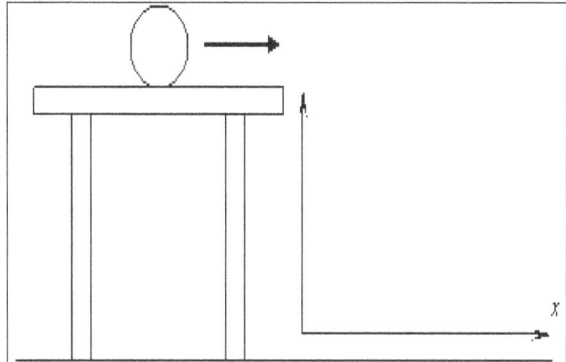

While the ball is on the table we observe that the initial x-component of velocity (vox) is 10 m/s (constant), the initial y-component of velocity is 0 m/s, the x-component of acceleration is 0 m/s2 and the y-component of acceleration is 0 m/s2. The components of acceleration and velocity are those parts of the velocity or acceleration that points in the x or y direction. Let us observe what happens the instant the ball leaves the table.

The initial velocity in the y-direction is still zero and the initial velocity in the x-direction remains 10 m/s. However, the ball is no longer in contact with the table and it falls freely. The gravitational acceleration of the ball is down. In this case, the motions in the horizontal and vertical directions should be analyzed independently. Horizontally, there is no acceleration in the horizontal direction, therefore, the x-component of velocity is constant:

$$a_x = 0$$
$$v_x = \text{constant}$$
$$x = v_{0x'}$$

In the vertical direction there is an acceleration equal to the acceleration of gravity. Therefore, the velocity in the vertical direction changes as below:

$$a_y = g$$
$$v_y = v_{0y} + gt$$

Rotational Motion

Rotational motion deals only with rigid bodies. A rigid body is an object that retains its overall shape, meaning that the particles that make up the rigid body remain in the same position relative to one another. A wheel and rotor of a motor are common examples of rigid bodies that commonly appear in questions involving rotational motion.

Circular Motion

Circular motion is a common type of rotational motion. Like projectile motion we can analyze the kinematics and learn something about the relationships between position, velocity and acceleration. Newton's first law states that an object in motion remains in motion at constant velocity unless acted upon by an outside force. If the force is applied perpendicular to the direction of motion, only the direction of velocity will change. If a force constantly acts perpendicular to a moving object, the object will move in a circular path at constant speed. This is called uniform circular motion.

The circular motion of a rigid body occurs when every point in the body moves in a circular path around a line called the axis of rotation, which cuts through the center of mass as shown in the following figure.

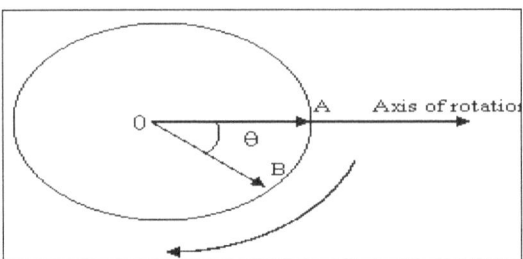

Uniform Circular Motion

An online simulation to measure the position, velocity, and acceleration (both components and magnitude) of an object undergoing circular motion.

OSCILLATORY MOTION

Oscillatory motion is defined as the to and fro motion of an object from its mean position. The ideal condition is that the object can be in oscillatory motion forever in the absence of friction but in the real world, this is not possible and the object has to settle into equilibrium.

To describe mechanical oscillation, the term vibration is used which is found in a swinging pendulum. Likewise, the beating of the human heart is an example of oscillation in dynamic systems.

Examples of Oscillatory Motion

Following are the examples of oscillatory motion:
- Oscillation of simple pendulum.
- Vibrating strings of musical instruments is a mechanical example of oscillatory motion.
- Movement of spring.
- Alternating current is an electrical example of oscillatory motion.
- Series of oscillations are seen in cosmological model.

Simple Harmonic Motion

Simple harmonic motion (SHM) is a type of oscillatory motion which is defined for the particle moving along the straight line with an acceleration which is moving towards a fixed point on the line such that the magnitude is proportional to the distance from the fixed point.

For any simple mechanical harmonic system (system of the weight hung by the spring to the wall) that is displaced from its equilibrium position, a restoring force which obeys the Hooke's law is required to restore the system back to equilibrium. Following is the mathematical representation of restoring force:

$$F=-kx$$

Where,

- F: restoring elastic force exerted by the spring (N).

- k: spring constant (Nm-1).

- x: displacement from equilibrium position (m).

Difference between Periodic Motion and Oscillatory Motion

Periodic motion is defined as the motion that repeats itself after fixed intervals of time. This fixed interval of time is known as time period of the periodic motion. Examples of periodic motion are motion of hands of the clock, motion of planets around the sun etc.

Oscillatory motion is defined as the to and fro motion of the body about its fixed position. Oscillatory motion is a type of periodic motion. Examples of oscillatory motion are vibrating strings, swinging of the swing etc.

NEWTON'S LAWS OF MOTION

Newton's laws of motion are the relations between the forces acting on a body and the motion of the body, first formulated by English physicist and mathematician Sir Isaac Newton.

Newton's first law states that, if a body is at rest or moving at a constant speed in a straight line, it will remain at rest or keep moving in a straight line at constant speed unless it is acted upon by a force. This postulate is known as the law of inertia. The law of inertia was first formulated by Galileo Galilei for horizontal motion on Earth and was later generalized by René Descartes. Before Galileo it had been thought that all horizontal motion required a direct cause, but Galileo deduced from his experiments that a body in motion would remain in motion unless a force (such as friction) caused it to come to rest.

Newton's second law is a quantitative description of the changes that a force can produce on the motion of a body. It states that the time rate of change of the momentum of a body is equal in both magnitude and direction to the force imposed on it. The momentum of a body is equal to the product of its mass and its velocity. Momentum, like velocity, is a vector quantity, having both magnitude and direction. A force applied to a body can change the magnitude of the momentum, or its direction, or both. Newton's second law is one of the most important in all of physics. For a body whose mass m is constant, it can be written in the form F = ma, where F (force) and a (acceleration) are both vector quantities. If a body has a net force acting on it, it is accelerated in accordance with the equation. Conversely, if a body is not accelerated, there is no net force acting on it.

Basketball; Newton's laws of motion. When a basketball player shoots a jump shot,
the ball always follows an arcing path. The ball follows this path because
its motion obeys Sir Isaac Newton's laws of motion.

Newton's third law states that when two bodies interact, they apply forces to one another that are equal in magnitude and opposite in direction. The third law is also known as the law of action and reaction. This law is important in analyzing problems of static equilibrium, where all forces are balanced, but it also applies to bodies in uniform or accelerated motion. The forces it describes are real ones, not mere bookkeeping devices. For example, a book resting on a table applies a downward force equal to its weight on the table. According to the third law, the table applies an equal and opposite force to the book. This force occurs because the weight of the book causes the table to deform slightly so that it pushes back on the book like a coiled spring.

Newton's laws first appeared in his masterpiece, Philosophiae Naturalis Principia Mathematica, commonly known as the Principia. In 1543 Nicolaus Copernicus suggested that the Sun, rather than Earth, might be at the centre of the universe. In the intervening years Galileo, Johannes Kepler, and Descartes laid the foundations of a new science that would both replace the Aristotelian worldview, inherited from the ancient Greeks, and explain the workings of a heliocentric universe. In the Principia Newton created that new science. He developed his three laws in order to explain why the orbits of the planets are ellipses rather than circles, at which he succeeded, but it turned out that he explained much more. The series of events from Copernicus to Newton is known collectively as the scientific revolution.

In the 20th century Newton's laws were replaced by quantum mechanics and relativity as the most fundamental laws of physics. Nevertheless, Newton's laws continue to give an accurate account of nature, except for very small bodies such as electrons or for bodies moving close to the speed of light. Quantum mechanics and relativity reduce to Newton's laws for larger bodies or for bodies moving more slowly.

ENERGY

Energy is a basic necessity of life. The sustenance and quality of life depend on its availability. Hence, it is important for us to understand what energy is. In this topic, we discuss about energy and its different types along with the law of conservation of energy.

We can define energy as the strength to do any kind of physical activity. Thus, they say, " Energy is the ability to do work ". Physical or chemical resources are processed to generate energy which is further used to provide light or heat for domestic or industrial purposes. We have also heard people comparing two persons (A & B) and concluding that A has more energy than B. Thus, we can conclude that various types of energy which can never be created nor destroyed. Energy can only be transformed from one form to another.

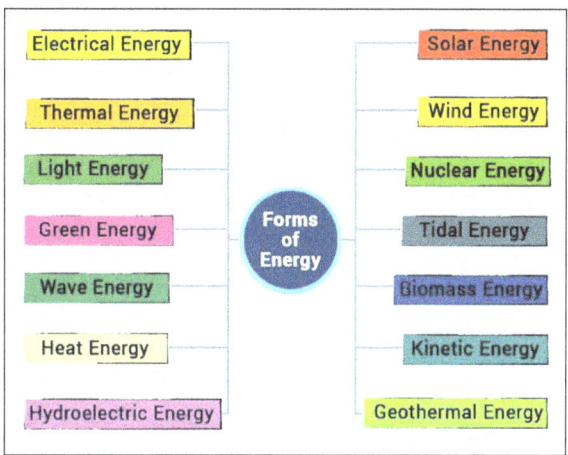

Unit of Energy

The SI unit of energy is the same as that of work, i.e. Joules (J) which is nothing but a term for newton-meter. When a certain amount of force (Newton) is applied to an object and it moved a certain distance (meters), then the energy applied is said to be Joules (newton-meters).

Different Types of Energy

There are different forms of energy but the distinction between them is not always clear. As Richard Feynman, a famous physicist once said, "The notions of potential and kinetic energy depend on a notion of length scale. For example, one can speak of macroscopic potential and kinetic energy, which do not include thermal potential and kinetic energy. Also what is called chemical potential energy is a macroscopic notion, and closer examination shows that it is really the sum of the potential and kinetic energy on the atomic and subatomic scale. Similar remarks apply to nuclear "potential" energy and most other forms of energy."

Some important types of energy and their features:

- Kinetic Energy

The energy in motion is known as Kinetic Energy. For example a moving ball, flowing water etc.

$$KineticEnergy = \frac{1}{2} m \times v^2$$

Where,

      m = Mass of the object

      v = Velocity of the object

- Potential Energy

This is the energy stored in an object and is measured by the amount of work done. For example, a pen on a table, water in a lake etc.

Potential Energy = m×g×h

Where,

m = Mass of the object (in kilograms)

g = Acceleration due to gravity

h = Height in meters

- Mechanical Energy

It is the sum of potential energy and kinetic energy that is the energy associated with the motion & position of an object is known as Mechanical energy. Thus, we can derive the formula of mechanical energy as:

Mechanical Energy = Kinetic Energy + Potential Energy

Mechanical Energy$= \frac{1}{2}m \times v^2 + m \times g \times h$

- Solar Energy

The light and heat from the sun, harnessed using technologies like, solar heating, photovoltaics, solar thermal energy, solar architecture and artificial photosynthesis is known as solar energy. It is the prime source of renewable-energy.

- Wind Energy

It is one of the various forms of energy. The energy present in the flow of wind, used by wind turbines is called the wind energy. This energy is a major cheap source to produce electricity. In this phenomena, the kinetic energy of the wind is converted into mechanical power.

- Nuclear Energy

The energy present in the nucleus of an atom is known as nuclear energy. The particles of an atom are tiny and need energy to hold themselves. Nuclear energy is that enormous energy in the bonds of an atom which helps to hold the atom together. Nuclear energy can be used to make electricity.

- Geothermal Energy

The energy or heat present inside the Earth is known as geothermal energy. It is a cheap & convenient heat and power resource and use of this energy doesn't have a side effect like greenhouse gas emission etc.

- Tidal Energy

Tidal energy or tidal power is a form of hydropower (energy present in water), which converts the energy present in the tides to produce electricity.

- Biomass Energy

Biomass is organic matter obtained from living organisms. The energy produced from biomass is called biomass energy.

- Electrical Energy

The energy caused by moving electric charges is known as electrical energy. Electric energy is a type of kinetic energy as the electrical charges moves.

- Thermal Energy

As the name suggests, thermal energy is the energy obtained from heat. It is a microscopic, disordered equivalent of mechanical energy.

There may be instances where an object posses more than one type of energy. For example, boiling water, posses both kinetic and potential energy along with heat energy.

Law of Conservation of Energy

The law of conservation of energy is one of the basic laws in physics. It governs the microscopic motion of individual atoms in a chemical reaction. The law of conservation of energy states that *"In a closed system, i.e., a system that isolated from its surroundings, the total energy of the system is conserved."* According to the law, the total energy in a system is conserved even though the transformation of energy occurs. Energy can neither be created nor destroyed, it can only be converted from one form to another.

The understanding and development of energy are crucial for societal development. Our ability to utilise energy effectively improves the quality of life. It is hard to imagine a life without energy.

POTENTIAL ENERGY

In physics, potential energy is the energy held by an object because of its position relative to other objects, stresses within itself, its electric charge, or other factors.

Common types of potential energy include the gravitational potential energy of an object that depends on its mass and its distance from the center of mass of another object, the elastic potential energy of an extended spring, and the electric potential energy of an electric charge in an electric field. The unit for energy in the International System of Units (SI) is the joule, which has the symbol J.

The term *potential energy* was introduced by the 19th-century Scottish engineer and physicist William Rankine, although it has links to Greek philosopher Aristotle's concept of potentiality. Potential energy is associated with forces that act on a body in a way that the total work done by these forces on the body depends only on the initial and final positions of the body in space. These forces, that are called *conservative forces*, can be represented at every point in space by vectors expressed as gradients of a certain scalar function called *potential*.

Since the work of potential forces acting on a body that moves from a start to an end position is determined only by these two positions, and does not depend on the trajectory of the body, there is a function known as *potential* that can be evaluated at the two positions to determine this work.

There are various types of potential energy, each associated with a particular type of force. For example, the work of an elastic force is called elastic potential energy; work of the gravitational force is called gravitational potential energy; work of the Coulomb force is called electric potential energy; work of the strong nuclear force or weak nuclear force acting on the baryon charge is called nuclear potential energy; work of intermolecular forces is called intermolecular potential energy. Chemical potential energy, such as the energy stored in fossil fuels, is the work of the Coulomb force during rearrangement of mutual positions of electrons and nuclei in atoms and molecules. Thermal energy usually has two components: the kinetic energy of random motions of particles and the potential energy of their mutual positions.

Forces derivable from a potential are also called conservative forces. The work done by a conservative force is:

$$W = -\Delta U$$

where ΔU is the change in the potential energy associated with the force. The negative sign provides the convention that work done against a force field increases potential energy, while work done by the force field decreases potential energy. Common notations for potential energy are PE, U, V, and E_p.

Potential energy is the energy by virtue of an object's position relative to other objects. Potential energy is often associated with restoring forces such as a spring or the force of gravity. The action of stretching a spring or lifting a mass is performed by an external force that works against the force field of the potential. This work is stored in the force field, which is said to be stored as potential energy. If the external force is removed the force field acts on the body to perform the work as it moves the body back to the initial position, reducing the stretch of the spring or causing a body to fall.

Consider a ball whose mass is m and whose height is h. The acceleration g of free fall is approximately constant, so the weight force of the ball mg is constant. Force × displacement gives the work done, which is equal to the gravitational potential energy, thus:

$$U_g = mgh$$

The more formal definition is that potential energy is the energy difference between the energy of an object in a given position and its energy at a reference position.

Work and Potential Energy

Potential energy is closely linked with forces. If the work done by a force on a body that moves from A to B does not depend on the path between these points (if the work is done by a conservative force), then the work of this force measured from A assigns a scalar value to every other point in space and defines a scalar potential field. In this case, the force can be defined as the negative of the vector gradient of the potential field.

If the work for an applied force is independent of the path, then the work done by the force is evaluated at the start and end of the trajectory of the point of application. This means that there is a function $U(\mathbf{x})$, called a "potential," that can be evaluated at the two points $\mathbf{x}_A$ and $\mathbf{x}_B$ to obtain the work over any trajectory between these two points. It is tradition to define this function with a negative sign so that positive work is a reduction in the potential, that is:

$$W = \int_C \mathbf{F} \cdot d\mathbf{x} = U(\mathbf{x}_A) - U(\mathbf{x}_B)$$

where C is the trajectory taken from A to B. Because the work done is independent of the path taken, then this expression is true for any trajectory, C, from A to B.

The function $U(\mathbf{x})$ is called the potential energy associated with the applied force. Examples of forces that have potential energies are gravity and spring forces.

Derivable from a Potential

The line integral that defines work along curve C takes a special form if the force $\mathbf{F}$ is related to a scalar field $\varphi(\mathbf{x})$ so that:

$$\mathbf{F} = \nabla\varphi = \left(\frac{\partial\varphi}{\partial x}, \frac{\partial\varphi}{\partial y}, \frac{\partial\varphi}{\partial z} \right).$$

In this case, work along the curve is given by:

$$W = \int_C \mathbf{F} \cdot d\mathbf{x} = \int_C \nabla\varphi \cdot d\mathbf{x},$$

which can be evaluated using the gradient theorem to obtain:

$$W = \varphi(\mathbf{x}_B) - \varphi(\mathbf{x}_A).$$

This shows that when forces are derivable from a scalar field, the work of those forces along a curve C is computed by evaluating the scalar field at the start point A and the end point B of the curve. This means the work integral does not depend on the path between A and B and is said to be independent of the path.

Potential energy $U=-\varphi(\mathbf{x})$ is traditionally defined as the negative of this scalar field so that work by the force field decreases potential energy, that is:

$$W = U(\mathbf{x}_A) - U(\mathbf{x}_B).$$

In this case, the application of the del operator to the work function yields,

$$\nabla W = -\nabla U = -\left(\frac{\partial U}{\partial x}, \frac{\partial U}{\partial y}, \frac{\partial U}{\partial z} \right) = \mathbf{F},$$

and the force F is said to be "derivable from a potential." This also necessarily implies that F must be a conservative vector field. The potential U defines a force $\mathbf{F}$ at every point $\mathbf{x}$ in space, so the set of forces is called a force field.

Computing Potential Energy

Given a force field $\mathbf{F}(\mathbf{x})$, evaluation of the work integral using the gradient theorem can be used to find the scalar function associated with potential energy. This is done by introducing a parameterized curve $\gamma(t)=\mathbf{r}(t)$ from $\gamma(a)=A$ to $\gamma(b)=B$, and computing,

$$
\begin{aligned}
\int_{\gamma} \nabla\varphi(\mathbf{r}) \cdot d\mathbf{r} &= \int_{a}^{b} \nabla\varphi(\mathbf{r}(t)) \cdot \mathbf{r}'(t)dt, \\
&= \int_{a}^{b} \frac{d}{dt}\varphi(\mathbf{r}(t))dt = \varphi(\mathbf{r}(b)) - \varphi(\mathbf{r}(a)) = \varphi(\mathbf{x}_{B}) - \varphi(\mathbf{x}_{A}).
\end{aligned}
$$

For the force field $\mathbf{F}$, let $\mathbf{v} = d\mathbf{r}/dt$, then the gradient theorem yields,

$$
\begin{aligned}
\int_{\gamma} \mathbf{F} \cdot d\mathbf{r} &= \int_{a}^{b} \mathbf{F} \cdot \mathbf{v}dt, \\
&= -\int_{a}^{b} \frac{d}{dt}U(\mathbf{r}(t))dt = U(\mathbf{x}_{A}) - U(\mathbf{x}_{B}).
\end{aligned}
$$

The power applied to a body by a force field is obtained from the gradient of the work, or potential, in the direction of the velocity $\mathbf{v}$ of the point of application, that is:

$$
P(t) = -\nabla U \cdot \mathbf{v} = \mathbf{F} \cdot \mathbf{v}.
$$

Examples of work that can be computed from potential functions are gravity and spring forces.

Potential Energy for Near Earth Gravity

A trebuchet uses the gravitational potential energy of the counterweight
to throw projectiles over two hundred meters.

For small height changes, gravitational potential energy can be computed using:

$$
U_{g} = mgh,
$$

where m is the mass in kg, g is the local gravitational field (9.8 metres per second squared on earth) and h is the height above a reference level in metres and U is the energy in joules.

In classical physics, gravity exerts a constant downward force $\mathbf{F}=(0, 0, F_{z})$ on the center of mass of a body moving near the surface of the Earth. The work of gravity on a body moving along a trajectory $r(t) = (x(t), y(t), z(t))$, such as the track of a roller coaster is calculated using its velocity, $v=(v_x, v_y, v_z)$, to obtain:

$$W = \int_{t_1}^{t_2} \mathbf{F} \cdot \mathbf{v} dt = \int_{t_1}^{t_2} F_z v_z dt = F_z \Delta z.$$

where the integral of the vertical component of velocity is the vertical distance. Notice that the work of gravity depends only on the vertical movement of the curve $r(t)$.

Potential Energy for a Linear Spring

Springs are used for storing elastic potential energy.

Archery is one of humankind's oldest applications of elastic potential energy.

A horizontal spring exerts a force $\mathbf{F} = (-kx, 0, 0)$ that is proportional to its deformation in the axial or x direction. The work of this spring on a body moving along the space curve $\mathbf{s}(t) = (x(t), y(t), z(t))$, is calculated using its velocity, $\mathbf{v} = (v_x, v_y, v_z)$, to obtain:

$$W = \int_0^t \mathbf{F} \cdot \mathbf{v} dt = -\int_0^t kxv_x dt = -\frac{1}{2}kx^2.$$

For convenience, consider contact with the spring occurs at $t = 0$, then the integral of the product of the distance x and the x-velocity, xv_x, is $x^2/2$.

The function:

$$U(x) = \frac{1}{2}kx^2,$$

is called the potential energy of a linear spring.

Elastic potential energy is the potential energy of an elastic object (for example a bow or a catapult) that is deformed under tension or compression (or stressed in formal terminology). It arises as a consequence of a force that tries to restore the object to its original shape, which is most often the electromagnetic force between the atoms and molecules that constitute the object. If the stretch is released, the energy is transformed into kinetic energy.

Potential Energy for Gravitational Forces between Two Bodies

The gravitational potential function, also known as gravitational potential energy, is:

$$U = -\frac{GMm}{r},$$

The negative sign follows the convention that work is gained from a loss of potential energy.

Derivation

The gravitational force between two bodies of mass M and m separated by a distance r is given by Newton's law:

$$\mathbf{F} = -\frac{GMm}{r^2}\hat{\mathbf{r}},$$

where $\hat{\mathbf{r}}$ is a vector of length 1 pointing from M to m and G is the gravitational constant.

Let the mass m move at the velocity $\mathbf{v}$ then the work of gravity on this mass as it moves from position $\mathbf{r}(t_1)$ to $\mathbf{r}(t_2)$ is given by:

$$W = -\int_{\mathbf{r}(t_1)}^{\mathbf{r}(t_2)} \frac{GMm}{r^3}\mathbf{r} \cdot d\mathbf{r} = -\int_{t_1}^{t_2} \frac{GMm}{r^3}\mathbf{r} \cdot \mathbf{v}dt.$$

Notice that the position and velocity of the mass m are given by:

$$\mathbf{r} = r\mathbf{e}_r, \qquad \mathbf{v} = \dot{r}\mathbf{e}_r + r\dot{\theta}\mathbf{e}_t,$$

where $\mathbf{e}_r$ and $\mathbf{e}_t$ are the radial and tangential unit vectors directed relative to the vector from M to m. Use this to simplify the formula for work of gravity to:

$$W = -\int_{t_1}^{t_2} \frac{GmM}{r^3}(r\mathbf{e}_r) \cdot (\dot{r}\mathbf{e}_r + r\dot{\theta}\mathbf{e}_t)dt = -\int_{t_1}^{t_2} \frac{GmM}{r^3}r\dot{r}dt = \frac{GMm}{r(t_2)} - \frac{GMm}{r(t_1)}.$$

This calculation uses the fact that:

$$\frac{d}{dt}r^{-1} = -r^{-2}\dot{r} = -\frac{\dot{r}}{r^2}.$$

Potential Energy for Electrostatic Forces between Two Bodies

The electrostatic force exerted by a charge Q on another charge q separated by a distance r is given by Coulomb's Law:

$$\mathbf{F} = \frac{1}{4\pi\varepsilon_0}\frac{Qq}{r^2}\hat{\mathbf{r}},$$

where $\hat{\mathbf{r}}$ is a vector of length 1 pointing from Q to q and ε_0 is the vacuum permittivity. This may also be written using Coulomb constant $k_e = 1/4\pi\varepsilon_0$.

The work W required to move q from A to any point B in the electrostatic force field is given by the potential function:

$$U(r) = \frac{1}{4\pi\varepsilon_0}\frac{Qq}{r}.$$

Reference Level

The potential energy is a function of the state a system is in, and is defined relative to that for a particular state. This reference state is not always a real state; it may also be a limit, such as with the distances between all bodies tending to infinity, provided that the energy involved in tending to that limit is finite, such as in the case of inverse-square law forces. Any arbitrary reference state could be used; therefore it can be chosen based on convenience.

Typically the potential energy of a system depends on the *relative* positions of its components only, so the reference state can also be expressed in terms of relative positions.

Gravitational Potential Energy

Gravitational energy is the potential energy associated with gravitational force, as work is required to elevate objects against Earth's gravity. The potential energy due to elevated positions is called gravitational potential energy, and is evidenced by water in an elevated reservoir or kept behind a dam. If an object falls from one point to another point inside a gravitational field, the force of gravity will do positive work on the object, and the gravitational potential energy will decrease by the same amount.

Gravitational force keeps the planets in orbit around the Sun.

Consider a book placed on top of a table. As the book is raised from the floor to the table, some external force works against the gravitational force. If the book falls back to the floor, the "falling" energy the book receives is provided by the gravitational force. Thus, if the book falls off the table, this potential energy goes to accelerate the mass of the book and is converted into kinetic energy. When the book hits the floor this kinetic energy is converted into heat, deformation, and sound by the impact.

The factors that affect an object's gravitational potential energy are its height relative to some reference point, its mass, and the strength of the gravitational field it is in. Thus, a book lying on a table has less gravitational potential energy than the same book on top of a taller cupboard and less gravitational potential energy than a heavier book lying on the same table. An object at a certain height above the Moon's surface has less gravitational potential energy than at the same height above the Earth's surface because the Moon's gravity is weaker. Note that "height" in the common sense of the term cannot be used for gravitational potential energy calculations when gravity is not assumed to be a constant. The following sections provide more detail.

Local Approximation

The strength of a gravitational field varies with location. However, when the change of distance is small in relation to the distances from the center of the source of the gravitational field, this variation in field strength is negligible and we can assume that the force of gravity on a particular object is constant. Near the surface of the Earth, for example, we assume that the acceleration due to gravity is a constant g = 9.8 m/s² ("standard gravity"). In this case, a simple expression for gravitational potential energy can be derived using the $W = Fd$ equation for work, and the equation:

$$W_F = -\Delta U_F.$$

The amount of gravitational potential energy held by an elevated object is equal to the work done against gravity in lifting it. The work done equals the force required to move it upward multiplied with the vertical distance it is moved (remember $W = Fd$). The upward force required while moving at a constant velocity is equal to the weight, mg, of an object, so the work done in lifting it through a height h is the product mgh. Thus, when accounting only for mass, gravity, and altitude, the equation is:

$$U = mgh$$

where U is the potential energy of the object relative to its being on the Earth's surface, m is the mass of the object, g is the acceleration due to gravity, and h is the altitude of the object. If m is expressed in kilograms, g in m/s² and h in metres then U will be calculated in joules.

Hence, the potential difference is:

$$\Delta U = mg\Delta h.$$

General Formula

However, over large variations in distance, the approximation that g is constant is no longer valid, and we have to use calculus and the general mathematical definition of work to determine gravitational potential energy. For the computation of the potential energy, we can integrate the gravitational force, whose magnitude is given by Newton's law of gravitation, with respect to the distance r between the two bodies. Using that definition, the gravitational potential energy of a system of masses m_1 and M_2 at a distance r using gravitational constant G is:

$$U = -G\frac{m_1 M_2}{r} + K$$,

where K is an arbitrary constant dependent on the choice of datum from which potential is measured. Choosing the convention that $K=0$ (i.e. in relation to a point at infinity) makes calculations simpler, albeit at the cost of making U negative; for why this is physically reasonable.

Given this formula for U, the total potential energy of a system of n bodies is found by summing, for all $\frac{n(n-1)}{2}$ pairs of two bodies, the potential energy of the system of those two bodies.

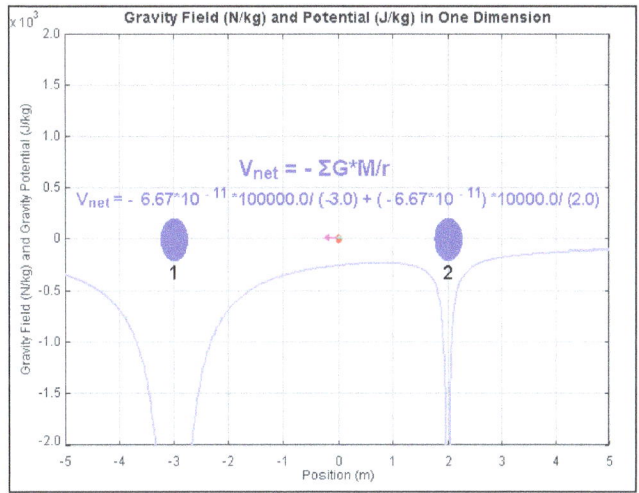

Gravitational potential summation $U = -m(G\dfrac{M_1}{r_1} + G\dfrac{M_2}{r_2})$.

Considering the system of bodies as the combined set of small particles the bodies consist of, and applying the previous on the particle level we get the negative gravitational binding energy. This potential energy is more strongly negative than the total potential energy of the system of bodies as such since it also includes the negative gravitational binding energy of each body. The potential energy of the system of bodies as such is the negative of the energy needed to separate the bodies from each other to infinity, while the gravitational binding energy is the energy needed to separate all particles from each other to infinity.

$$U = -m\left(G\frac{M_1}{r_1} + G\frac{M_2}{r_2} \right)$$

therefore,

$$U = -m\sum G\frac{M}{r},$$

Negative Gravitational Energy

As with all potential energies, only differences in gravitational potential energy matter for most physical purposes, and the choice of zero point is arbitrary. Given that there is no reasonable criterion for preferring one particular finite r over another, there seem to be only two reasonable choices for the distance at which U becomes zero: $r = 0$ and $r = \infty$. The choice of $U = 0$ at infinity may seem peculiar, and the consequence that gravitational energy is always negative may seem counterintuitive, but this choice allows gravitational potential energy values to be finite, albeit negative.

The singularity at $r = 0$ in the formula for gravitational potential energy means that the only other apparently reasonable alternative choice of convention, with $U = 0$ for $r = 0$, would result in potential energy being positive, but infinitely large for all nonzero values of r, and would make calculations involving sums or differences of potential energies beyond what is possible with the real number system. Since physicists abhor infinities in their calculations, and r is always non-zero in practice, the choice of $U = 0$ at infinity is by far the more preferable choice, even if the idea of negative energy in a gravity well appears to be peculiar at first.

The negative value for gravitational energy also has deeper implications that make it seem more reasonable in cosmological calculations where the total energy of the universe can meaningfully be considered.

Uses

Gravitational potential energy has a number of practical uses, notably the generation of pumped-storage hydroelectricity. For example, in Dinorwig, Wales, there are two lakes, one at a higher elevation than the other. At times when surplus electricity is not required (and so is comparatively cheap), water is pumped up to the higher lake, thus converting the electrical energy (running the pump) to gravitational potential energy. At times of peak demand for electricity, the water flows back down through electrical generator turbines, converting the potential energy into kinetic energy and then back into electricity. The process is not completely efficient and some of the original energy from the surplus electricity is in fact lost to friction.

Gravitational potential energy is also used to power clocks in which falling weights operate the mechanism.

It's also used by counterweights for lifting up an elevator, crane, or sash window.

Roller coasters are an entertaining way to utilize potential energy – chains are used to move a car up an incline (building up gravitational potential energy), to then have that energy converted into kinetic energy as it falls.

Another practical use is utilizing gravitational potential energy to descend (perhaps coast) downhill in transportation such as the descent of an automobile, truck, railroad train, bicycle, airplane, or fluid in a pipeline. In some cases the kinetic energy obtained from the potential energy of descent may be used to start ascending the next grade such as what happens when a road is undulating and has frequent dips. The commercialization of stored energy (in the form of rail cars raised to higher elevations) that is then converted to electrical energy when needed by an electrical grid, is being undertaken in the United States in a system called Advanced Rail Energy Storage (ARES).

Chemical Potential Energy

Chemical potential energy is a form of potential energy related to the structural arrangement of atoms or molecules. This arrangement may be the result of chemical bonds within a molecule or otherwise. Chemical energy of a chemical substance can be transformed to other forms of energy by a chemical reaction. As an example, when a fuel is burned the chemical energy is converted to heat, same is the case with digestion of food metabolized in a biological organism. Green plants transform solar energy to chemical energy through the process known as photosynthesis, and electrical energy can be converted to chemical energy through electrochemical reactions.

The similar term chemical potential is used to indicate the potential of a substance to undergo a change of configuration, be it in the form of a chemical reaction, spatial transport, particle exchange with a reservoir, etc.

Electric Potential Energy

An object can have potential energy by virtue of its electric charge and several forces related to their presence. There are two main types of this kind of potential energy: electrostatic potential energy, electrodynamic potential energy (also sometimes called magnetic potential energy).

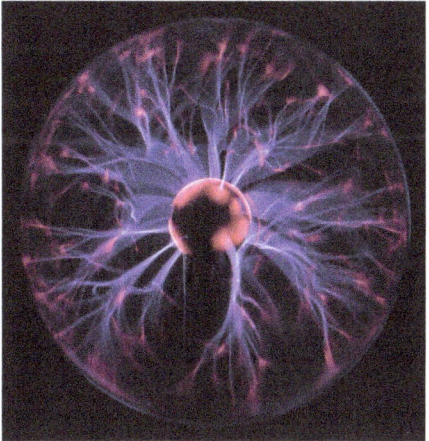
Plasma formed inside a gas filled sphere.

Electrostatic Potential Energy

Electrostatic potential energy between two bodies in space is obtained from the force exerted by a charge Q on another charge q which is given by:

$$\mathbf{F}_e = -\frac{1}{4\pi\varepsilon_0}\frac{Qq}{r^2}\hat{\mathbf{r}},$$

where $\hat{\mathbf{r}}$ is a vector of length 1 pointing from Q to q and ε_0 is the vacuum permittivity. This may also be written using Coulomb's constant $k_e = 1/4\pi\varepsilon_0$.

If the electric charge of an object can be assumed to be at rest, then it has potential energy due to its position relative to other charged objects. The electrostatic potential energy is the energy of an electrically charged particle (at rest) in an electric field. It is defined as the work that must be done to move it from an infinite distance away to its present location, adjusted for non-electrical forces on the object. This energy will generally be non-zero if there is another electrically charged object nearby.

The work W required to move q from A to any point B in the electrostatic force field is given by:

$$\Delta U_{AB}(\mathbf{r}) = -\int_A^B \mathbf{F_e} \cdot d\mathbf{r}$$

typically given in *J* for Joules. A related quantity called *electric potential* (commonly denoted with a *V* for voltage) is equal to the electric potential energy per unit charge.

Magnetic Potential Energy

The energy of a magnetic moment i in an externally produced magnetic B-field B has potential energy:

$$U = -\mu \cdot B.$$

The magnetization M in a field is:

$$U = -\frac{1}{2}\int \mathbf{M} \cdot \mathbf{B} dV,$$

where the integral can be over all space or, equivalently, where **M** is nonzero. Magnetic potential energy is the form of energy related not only to the distance between magnetic materials, but also to the orientation, or alignment, of those materials within the field. For example, the needle of a compass has the lowest magnetic potential energy when it is aligned with the north and south poles of the Earth's magnetic field. If the needle is moved by an outside force, torque is exerted on the magnetic dipole of the needle by the Earth's magnetic field, causing it to move back into alignment. The magnetic potential energy of the needle is highest when its field is in the same direction as the Earth's magnetic field. Two magnets will have potential energy in relation to each other and the distance between them, but this also depends on their orientation. If the opposite poles are held apart, the potential energy will be higher the further they are apart and lower the closer they are. Conversely, like poles will have the highest potential energy when forced together, and the lowest when they spring apart.

Nuclear Potential Energy

Nuclear potential energy is the potential energy of the particles inside an atomic nucleus. The nuclear particles are bound together by the strong nuclear force. Weak nuclear forces provide the potential energy for certain kinds of radioactive decay, such as beta decay.

Nuclear particles like protons and neutrons are not destroyed in fission and fusion processes, but collections of them can have less mass than if they were individually free, in which case this mass difference can be liberated as heat and radiation in nuclear reactions (the heat and radiation have the missing mass, but it often escapes from the system, where it is not measured). The energy from the Sun is an example of this form of energy conversion. In the Sun, the process of hydrogen fusion converts about 4 million tonnes of solar matter per second into electromagnetic energy, which is radiated into space.

Forces and Potential Energy

Potential energy is closely linked with forces. If the work done by a force on a body that moves from *A* to *B* does not depend on the path between these points, then the work of this force measured from *A* assigns a scalar value to every other point in space and defines a scalar potential field. In this case, the force can be defined as the negative of the vector gradient of the potential field.

For example, gravity is a conservative force. The associated potential is the gravitational potential, often denoted by ϕ or V, corresponding to the energy per unit mass as a function of position. The gravitational potential energy of two particles of mass M and m separated by a distance r is:

$$U = -\frac{GMm}{r},$$

The gravitational potential (specific energy) of the two bodies is:

$$\phi = -\left(\frac{GM}{r} + \frac{Gm}{r}\right) = -\frac{G(M+m)}{r} = -\frac{GMm}{\mu r} = \frac{U}{\mu}.$$

where μ is the reduced mass.

The work done against gravity by moving an infinitesimal mass from point A with $U = a$ to point B with $U = b$ is $(b - a)$ and the work done going back the other way is $(a - b)$ so that the total work done in moving from A to B and returning to A is:

$$U_{A \to B \to A} = (b - a) + (a - b) = 0.$$

If the potential is redefined at A to be $a + c$ and the potential at B to be $b + c$, where c is a constant (i.e. c can be any number, positive or negative, but it must be the same at A as it is at B) then the work done going from A to B is:

$$U_{A \to B} = (b + c) - (a + c) = b - a$$

as before.

In practical terms, this means that one can set the zero of U and ϕ anywhere one likes. One may set it to be zero at the surface of the Earth, or may find it more convenient to set zero at infinity.

A conservative force can be expressed in the language of differential geometry as a closed form. As Euclidean space is contractible, its de Rham cohomology vanishes, so every closed form is also an exact form, and can be expressed as the gradient of a scalar field. This gives a mathematical justification of the fact that all conservative forces are gradients of a potential field.

KINETIC ENERGY

In physics, the kinetic energy of an object is the energy that it possesses due to its motion. It is defined as the work needed to accelerate a body of a given mass from rest to its stated velocity. Having gained this energy during its acceleration, the body maintains this kinetic energy unless its speed changes. The same amount of work is done by the body when decelerating from its current speed to a state of rest.

In classical mechanics, the kinetic energy of a non-rotating object of mass m traveling at a speed v is $\frac{1}{2}mv^2$. In relativistic mechanics, this is a good approximation only when v is much less than the speed of light.

The standard unit of kinetic energy is the joule, while the imperial unit of kinetic energy is the foot-pound.

Energy occurs in many forms, including chemical energy, thermal energy, electromagnetic radiation, gravitational energy, electric energy, elastic energy, nuclear energy, and rest energy. These can be categorized in two main classes: potential energy and kinetic energy. Kinetic energy is the movement energy of an object. Kinetic energy can be transferred between objects and transformed into other kinds of energy.

Kinetic energy may be best understood by examples that demonstrate how it is transformed to and from other forms of energy. For example, a cyclist uses chemical energy provided by food to accelerate a bicycle to a chosen speed. On a level surface, this speed can be maintained without further work, except to overcome air resistance and friction. The chemical energy has been converted into kinetic energy, the energy of motion, but the process is not completely efficient and produces heat within the cyclist.

The kinetic energy in the moving cyclist and the bicycle can be converted to other forms. For example, the cyclist could encounter a hill just high enough to coast up, so that the bicycle comes to a complete halt at the top. The kinetic energy has now largely been converted to gravitational potential energy that can be released by freewheeling down the other side of the hill. Since the bicycle lost some of its energy to friction, it never regains all of its speed without additional pedaling. The energy is not destroyed; it has only been converted to another form by friction. Alternatively, the cyclist could connect a dynamo to one of the wheels and generate some electrical energy on the descent. The bicycle would be traveling slower at the bottom of the hill than without the generator because some of the energy has been diverted into electrical energy. Another possibility would be for the cyclist to apply the brakes, in which case the kinetic energy would be dissipated through friction as heat.

Like any physical quantity that is a function of velocity, the kinetic energy of an object depends on the relationship between the object and the observer's frame of reference. Thus, the kinetic energy of an object is not invariant.

Spacecraft use chemical energy to launch and gain considerable kinetic energy to reach orbital velocity. In an entirely circular orbit, this kinetic energy remains constant because there is almost no friction in near-earth space. However, it becomes apparent at re-entry when some of the kinetic energy is converted to heat. If the orbit is elliptical or hyperbolic, then throughout the orbit kinetic and potential energy are exchanged; kinetic energy is greatest and potential energy lowest at closest approach to the earth or other massive body, while potential energy is greatest and kinetic energy the lowest at maximum distance. Without loss or gain, however, the sum of the kinetic and potential energy remains constant.

Kinetic energy can be passed from one object to another. In the game of billiards, the player imposes kinetic energy on the cue ball by striking it with the cue stick. If the cue ball collides with another ball, it slows down dramatically, and the ball it hit accelerates its speed as the kinetic energy is passed on to it. Collisions in billiards are effectively elastic collisions, in which kinetic energy is preserved. In inelastic collisions, kinetic energy is dissipated in various forms of energy, such as heat, sound, binding energy (breaking bound structures).

Flywheels have been developed as a method of energy storage. This illustrates that kinetic energy is also stored in rotational motion.

Several mathematical descriptions of kinetic energy exist that describe it in the appropriate physical situation. For objects and processes in common human experience, the formula ½mv² given by Newtonian (classical) mechanics is suitable. However, if the speed of the object is comparable to the speed of light, relativistic effects become significant and the relativistic formula is used. If the object is on the atomic or sub-atomic scale, quantum mechanical effects are significant, and a quantum mechanical model must be employed.

Newtonian Kinetic Energy

Kinetic Energy of Rigid Bodies

In classical mechanics, the kinetic energy of a *point object* (an object so small that its mass can be assumed to exist at one point), or a non-rotating rigid body depends on the mass of the body as well as its speed. The kinetic energy is equal to 1/2 the product of the mass and the square of the speed. In formula form:

$$E_k = \tfrac{1}{2}mv^2$$

where m is the mass and v is the speed (or the velocity) of the body. In SI units, mass is measured in kilograms, speed in metres per second, and the resulting kinetic energy is in joules.

For example, one would calculate the kinetic energy of an 80 kg mass (about 180 lbs) traveling at 18 metres per second (about 40 mph, or 65 km/h) as:

$$E_k = \frac{1}{2} \cdot 80\text{kg} \cdot (18\text{m/s})^2 = 12,960\text{J} = 12.96\text{kJ}$$

When a person throws a ball, the person does work on it to give it speed as it leaves the hand. The moving ball can then hit something and push it, doing work on what it hits. The kinetic energy of a moving object is equal to the work required to bring it from rest to that speed, or the work the object can do while being brought to rest: net force × displacement = kinetic energy, i.e.,

$$Fs = \tfrac{1}{2}mv^2$$

Since the kinetic energy increases with the square of the speed, an object doubling its speed has four times as much kinetic energy. For example, a car traveling twice as fast as another requires four times as much distance to stop, assuming a constant braking force. As a consequence of this quadrupling, it takes four times the work to double the speed.

The kinetic energy of an object is related to its momentum by the equation:

$$E_k = \frac{p^2}{2m}$$

where,

p is momentum.

m is mass of the body.

For the *translational kinetic energy*, that is the kinetic energy associated with rectilinear motion, of a rigid body with constant mass m, whose center of mass is moving in a straight line with speed v, as seen above is equal to:

$$E_{\mathrm{t}} = \tfrac{1}{2} m v^2$$

where,

  m is the mass of the body.

  v is the speed of the center of mass of the body.

The kinetic energy of any entity depends on the reference frame in which it is measured. However the total energy of an isolated system, i.e. one in which energy can neither enter nor leave, does not change over time in the reference frame in which it is measured. Thus, the chemical energy converted to kinetic energy by a rocket engine is divided differently between the rocket ship and its exhaust stream depending upon the chosen reference frame. This is called the Oberth effect. But the total energy of the system, including kinetic energy, fuel chemical energy, heat, etc., is conserved over time, regardless of the choice of reference frame. Different observers moving with different reference frames would however disagree on the value of this conserved energy.

The kinetic energy of such systems depends on the choice of reference frame: the reference frame that gives the minimum value of that energy is the center of momentum frame, i.e. the reference frame in which the total momentum of the system is zero. This minimum kinetic energy contributes to the invariant mass of the system as a whole.

Derivation

The work done in accelerating a particle with mass m during the infinitesimal time interval dt is given by the dot product of *force* $\mathbf{F}$ and the infinitesimal *displacement* $d\mathbf{x}$:

$$\mathbf{F} \cdot d\mathbf{x} = \mathbf{F} \cdot \mathbf{v} dt = \frac{d\mathbf{p}}{dt} \cdot \mathbf{v} dt = \mathbf{v} \cdot d\mathbf{p} = \mathbf{v} \cdot d(m\mathbf{v}),$$

where we have assumed the relationship $\mathbf{p} = m\mathbf{v}$ and the validity of Newton's Second Law.

Applying the product rule we see that:

$$d(\mathbf{v} \cdot \mathbf{v}) = (d\mathbf{v}) \cdot \mathbf{v} + \mathbf{v} \cdot (d\mathbf{v}) = 2(\mathbf{v} \cdot d\mathbf{v}).$$

Therefore, (assuming constant mass so that $dm=0$), we have,

$$\mathbf{v} \cdot d(m\mathbf{v}) = \frac{m}{2} d(\mathbf{v} \cdot \mathbf{v}) = \frac{m}{2} dv^2 = d\left(\frac{mv^2}{2}\right).$$

Since this is a total differential (that is, it only depends on the final state, not how the particle got there), we can integrate it and call the result kinetic energy. Assuming the object was at rest at time

0, we integrate from time 0 to time t because the work done by the force to bring the object from rest to velocity v is equal to the work necessary to do the reverse:

$$E_k = \int_0^t \mathbf{F} \cdot d\mathbf{x} = \int_0^t \mathbf{v} \cdot d(m\mathbf{v}) = \int_0^v d\left(\frac{mv^2}{2}\right) = \frac{mv^2}{2}.$$

This equation states that the kinetic energy (E_k) is equal to the integral of the dot product of the velocity (**v**) of a body and the infinitesimal change of the body's momentum (**p**). It is assumed that the body starts with no kinetic energy when it is at rest (motionless).

Rotating Bodies

If a rigid body Q is rotating about any line through the center of mass then it has *rotational kinetic energy* (E_r) which is simply the sum of the kinetic energies of its moving parts, and is thus given by:

$$E_r = \int_Q \frac{v^2 dm}{2} = \int_Q \frac{(r\omega)^2 dm}{2} = \frac{\omega^2}{2} \int_Q r^2 dm = \frac{\omega^2}{2} I = \frac{1}{2} I \omega^2$$

where,

- ω is the body's angular velocity.

- r is the distance of any mass dm from that line.

- I is the body's moment of inertia, equal to $\int_Q r^2 dm$.

(In this equation the moment of inertia must be taken about an axis through the center of mass and the rotation measured by ω must be around that axis; more general equations exist for systems where the object is subject to wobble due to its eccentric shape).

Kinetic Energy of Systems

A system of bodies may have internal kinetic energy due to the relative motion of the bodies in the system. For example, in the Solar System the planets and planetoids are orbiting the Sun. In a tank of gas, the molecules are moving in all directions. The kinetic energy of the system is the sum of the kinetic energies of the bodies it contains.

A macroscopic body that is stationary (i.e. a reference frame has been chosen to correspond to the body's center of momentum) may have various kinds of internal energy at the molecular or atomic level, which may be regarded as kinetic energy, due to molecular translation, rotation, and vibration, electron translation and spin, and nuclear spin. These all contribute to the body's mass, as provided by the special theory of relativity. When discussing movements of a macroscopic body, the kinetic energy referred to is usually that of the macroscopic movement only. However all internal energies of all types contribute to body's mass, inertia, and total energy.

Fluid Dynamics

In fluid dynamics, the kinetic energy per unit volume at each point in an incompressible fluid flow field is called the dynamic pressure at that point.

$$E_k = \tfrac{1}{2}mv^2$$

Dividing by V, the unit of volume:

$$\frac{E_k}{V} = \tfrac{1}{2}\tfrac{m}{V}v^2$$

$$q = \tfrac{1}{2}\rho v^2$$

where q is the dynamic pressure, and ρ is the density of the incompressible fluid.

Frame of Reference

The speed, and thus the kinetic energy of a single object is frame-dependent (relative): it can take any non-negative value, by choosing a suitable inertial frame of reference. For example, a bullet passing an observer has kinetic energy in the reference frame of this observer. The same bullet is stationary to an observer moving with the same velocity as the bullet, and so has zero kinetic energy. By contrast, the total kinetic energy of a system of objects cannot be reduced to zero by a suitable choice of the inertial reference frame, unless all the objects have the same velocity. In any other case, the total kinetic energy has a non-zero minimum, as no inertial reference frame can be chosen in which all the objects are stationary. This minimum kinetic energy contributes to the system's invariant mass, which is independent of the reference frame.

The total kinetic energy of a system depends on the inertial frame of reference: it is the sum of the total kinetic energy in a center of momentum frame and the kinetic energy the total mass would have if it were concentrated in the center of mass.

This may be simply shown: let $\mathbf{V}$ be the relative velocity of the center of mass frame i in the frame k. Since $v^2 = (v_i + V)^2 = (\mathbf{v}_i + \mathbf{V})\cdot(\mathbf{v}_i + \mathbf{V}) = \mathbf{v}_i \cdot \mathbf{v}_i + 2\mathbf{v}_i \cdot \mathbf{V} + \mathbf{V}\cdot\mathbf{V} = v_i^2 + 2\mathbf{v}_i \cdot \mathbf{V} + V^2$, than:

$$E_k = \int \frac{v^2}{2}dm = \int \frac{v_i^2}{2}dm + \mathbf{V}\cdot\int \mathbf{v}_i dm + \frac{V^2}{2}\int dm.$$

However, let $\int \frac{v_i^2}{2}dm = E_i$ the kinetic energy in the center of mass frame, $\int \mathbf{v}_i dm$ would be simply the total momentum that is by definition zero in the center of mass frame, and let the total mass: $\int dm = M$. Substituting, we get:

$$E_k = E_i + \frac{MV^2}{2}.$$

Thus the kinetic energy of a system is lowest to center of momentum reference frames, i.e., frames of reference in which the center of mass is stationary (either the center of mass frame or any other center of momentum frame). In any different frame of reference, there is additional kinetic energy corresponding to the total mass moving at the speed of the center of mass. The kinetic energy of the system in the center of momentum frame is a quantity that is invariant.

Rotation in Systems

It sometimes is convenient to split the total kinetic energy of a body into the sum of the body's center-of-mass translational kinetic energy and the energy of rotation around the center of mass (rotational energy):

$$E_k = E_t + E_r$$

where:

E_k is the total kinetic energy.

E_t is the translational kinetic energy.

E_r is the *rotational energy* or *angular kinetic energy* in the rest frame.

Thus the kinetic energy of a tennis ball in flight is the kinetic energy due to its rotation, plus the kinetic energy due to its translation.

Relativistic Kinetic Energy of Rigid Bodies

If a body's speed is a significant fraction of the speed of light, it is necessary to use relativistic mechanics to calculate its kinetic energy. In special relativity theory, the expression for linear momentum is modified.

With m being an object's rest mass, $\mathbf{v}$ and v its velocity and speed, and c the speed of light in vacuum, we use the expression for linear momentum $\mathbf{p} = m\gamma\mathbf{v}$, where $\gamma = 1/\sqrt{1 - v^2/c^2}$.

Integrating by parts yields:

$$E_k = \int \mathbf{v} \cdot d\mathbf{p} = \int \mathbf{v} \cdot d(m\gamma\mathbf{v}) = m\gamma\mathbf{v} \cdot \mathbf{v} - \int m\gamma\mathbf{v} \cdot d\mathbf{v} = m\gamma v^2 - \frac{m}{2}\int \gamma d(v^2)$$

Since $\gamma = (1 - v^2/c^2)^{-1/2}$,

$$E_k = m\gamma v^2 - \frac{-mc^2}{2}\int \gamma d(1 - v^2/c^2)$$
$$= m\gamma v^2 + mc^2(1 - v^2/c^2)^{1/2} - E_0$$

E_0 is a constant of integration for the indefinite integral. Simplifying the expression we obtain:

$$E_k = m\gamma(v^2 + c^2(1 - v^2/c^2)) - E_0$$
$$= m\gamma(v^2 + c^2 - v^2) - E_0$$
$$= m\gamma c^2 - E_0$$

E_0 is found by observing that when $\mathbf{v} = 0, \gamma = 1$ and $E_k = 0$, giving:

$$E_0 = mc^2$$

resulting in the formula:

$$E = m\ c^2 - mc^2 = \frac{mc}{\sqrt{1\ v^2/c^2}} - mc^2$$

This formula shows that the work expended accelerating an object from rest approaches infinity as the velocity approaches the speed of light. Thus it is impossible to accelerate an object across this boundary.

The mathematical by-product of this calculation is the mass-energy equivalence formula—the body at rest must have energy content:

$$E_{rest} = E_0 = mc^2$$

At a low speed ($v << c$), the relativistic kinetic energy is approximated well by the classical kinetic energy. This is done by binomial approximation or by taking the first two terms of the Taylor expansion for the reciprocal square root:

$$E_k \approx mc^2 \left(1 + \frac{1}{2}v^2/c^2\right) - mc^2 = \frac{1}{2}mv^2$$

So, the total energy E_k can be partitioned into the rest mass energy plus the Newtonian kinetic energy at low speeds.

When objects move at a speed much slower than light (e.g. in everyday phenomena on Earth), the first two terms of the series predominate. The next term in the Taylor series approximation:

$$E_k \approx mc^2 \left(1 + \frac{1}{2}v^2/c^2 + \frac{3}{8}v^4/c^4\right) - mc^2 = \frac{1}{2}mv^2 + \frac{3}{8}mv^4/c^2$$

is small for low speeds. For example, for a speed of 10 km/s (22,000 mph) the correction to the Newtonian kinetic energy is 0.0417 J/kg (on a Newtonian kinetic energy of 50 MJ/kg) and for a speed of 100 km/s it is 417 J/kg (on a Newtonian kinetic energy of 5 GJ/kg).

The relativistic relation between kinetic energy and momentum is given by:

$$E_k = \sqrt{p^2c^2 + m^2c^4} - mc^2$$

This can also be expanded as a Taylor series, the first term of which is the simple expression from Newtonian mechanics:

$$E_k \approx \frac{p^2}{2m} - \frac{p^4}{8m^3c^2}$$

This suggests that the formulae for energy and momentum are not special and axiomatic, but concepts emerging from the equivalence of mass and energy and the principles of relativity.

General Relativity

Using the convention that:

$$g_{\alpha\beta}u^\alpha u^\beta = -c^2$$

where the four-velocity of a particle is:

$$u^\alpha = \frac{dx^\alpha}{d\tau}$$

and τ is the proper time of the particle, there is also an expression for the kinetic energy of the particle in general relativity.

If the particle has momentum:

$$p_\beta = m g_{\beta\alpha} u^\alpha$$

as it passes by an observer with four-velocity u_{obs}, then the expression for total energy of the particle as observed (measured in a local inertial frame) is:

$$E = -p_\beta u^\beta_{obs}$$

and the kinetic energy can be expressed as the total energy minus the rest energy:

$$E_k = -p_\beta u^\beta_{obs} - mc^2.$$

Consider the case of a metric that is diagonal and spatially isotropic $(g_{tt}, g_{ss}, g_{ss}, g_{ss})$. Since:

$$u^\alpha = \frac{dx^\alpha}{dt} \frac{dt}{d\tau} = v^\alpha u^t$$

where v^α is the ordinary velocity measured w.r.t. the coordinate system, we get:

$$-c^2 = g_{\alpha\beta} u^\alpha u^\beta = g_{tt}(u^t)^2 + g_{ss} v^2 (u^t)^2.$$

Solving for u^t gives:

$$u^t = c \sqrt{\frac{-1}{g_{tt} + g_{ss} v^2}}.$$

Thus for a stationary observer $(v = 0)$:

$$u^t_{obs} = c \sqrt{\frac{-1}{g_{tt}}}$$

and thus the kinetic energy takes the form:

$$E_k = -m g_{tt} u^t u^t_{obs} - mc^2 = mc^2 \sqrt{\frac{g_{tt}}{g_{tt} + g_{ss} v^2}} - mc^2.$$

Factoring out the rest energy gives:

$$E_k = mc^2 \left(\sqrt{\frac{g_{tt}}{g_{tt} + g_{ss} v^2}} - 1 \right).$$

This expression reduces to the special relativistic case for the flat-space metric where,

$$g_{tt} = -c^2$$

$$g_{ss} = 1.$$

In the Newtonian approximation to general relativity:

$$g_{tt} = -\left(c^2 + 2\Phi\right)$$

$$g_{ss} = 1 - \frac{2\Phi}{c^2}$$

where Φ is the Newtonian gravitational potential. This means clocks run slower and measuring rods are shorter near massive bodies.

Kinetic Energy in Quantum Mechanics

In quantum mechanics, observables like kinetic energy are represented as operators. For one particle of mass m, the kinetic energy operator appears as a term in the Hamiltonian and is defined in terms of the more fundamental momentum operator $\hat{p}$. The kinetic energy operator in the non-relativistic case can be written as:

$$\hat{T} = \frac{\hat{p}^2}{2m}.$$

Notice that this can be obtained by replacing p by $\hat{p}$ in the classical expression for kinetic energy in terms of momentum,

$$E_k = \frac{p^2}{2m}.$$

In the Schrödinger picture, $\hat{p}$ takes the form $-i\hbar\nabla$ where the derivative is taken with respect to position coordinates and hence:

$$\hat{T} = -\frac{\hbar^2}{2m}\nabla^2.$$

The expectation value of the electron kinetic energy, $\langle \hat{T} \rangle$, for a system of N electrons described by the wavefunction $|\psi\rangle$ is a sum of 1-electron operator expectation values:

$$\langle \hat{T} \rangle = \left\langle \psi \,\middle|\, \sum_{i=1}^{N} \frac{-\hbar^2}{2m_e} \nabla_i^2 \,\middle|\, \psi \right\rangle = -\frac{\hbar^2}{2m_e} \sum_{i=1}^{N} \left\langle \psi \,\middle|\, \nabla_i^2 \,\middle|\, \psi \right\rangle$$

where ∇_i^2 is the mass of the electron and ∇_i^2 is the Laplacian operator acting upon the coordinates of the ith electron and the summation runs over all electrons.

The density functional formalism of quantum mechanics requires knowledge of the electron density *only*, i.e., it formally does not require knowledge of the wavefunction. Given an electron density

$\rho(\mathbf{r})$, the exact N-electron kinetic energy functional is unknown; however, for the specific case of a 1-electron system, the kinetic energy can be written as:

$$T[\rho] = \frac{1}{8} \int \frac{\nabla \rho(\mathbf{r}) \cdot \nabla \rho(\mathbf{r})}{\rho(\mathbf{r})} d^3 r$$

where $T[\rho]$ is known as the von Weizsäcker kinetic energy functional.

References

- Force-and-its-effects, force-and-pressure, physics, guides: toppr.com, Retrieved 22 February, 2019

- Tipler, Paul (2004). Physics for Scientists and Engineers: Mechanics, Oscillations and Waves, Thermodynamics (5th ed.). W. H. Freeman. ISBN 0-7167-0809-4

- Friction, friction: britannica.com, Retrieved 23 March, 2019

- Motion, physics; physics-and-radio-electronics.com, Retrieved 24 April, 2019

- School of Mathematics and Statistics, University of St Andrews (2000). "Biography of Gaspard-Gustave de Coriolis (1792-1843)". Retrieved 2006-03-03

- Translational, resources: g9toengineering.com, Retrieved 25 May, 2019

- Oscillatory-motion, physics: byjus.com, Retrieved 26 June, 2019

- Newtons-laws-of-motion, science: britannica.com, Retrieved 27 July, 2019

- Tipler, Paul (2004). Physics for Scientists and Engineers: Mechanics, Oscillations and Waves, Thermodynamics (5th ed.). W. H. Freeman. ISBN 0-7167-0809-4

- Energy, physics: byjus.com, Retrieved 28 August, 2019

3

Light

Light is a form of electromagnetic radiation which requires no material medium to propagate. It illuminates any object on which it falls. The three main phenomena that form laws of light are reflection, refraction and diffraction of light. All these diverse principles of light have been carefully analysed in this chapter.

Light is a type of electromagnetic radiation that can be detected by the human eye. Electromagnetic radiation occurs over an extremely wide range of wavelengths, from gamma rays with wavelengths less than about 1×10^{-11} metre to radio waves measured in metres. Within that broad spectrum the wavelengths visible to humans occupy a very narrow band, from about 700 nanometres (nm; billionths of a metre) for red light down to about 400 nm for violet light. The spectral regions adjacent to the visible band are often referred to as light also, infrared at the one end and ultraviolet at the other. The speed of light in a vacuum is a fundamental physical constant, the currently accepted value of which is exactly 299,792,458 metres per second, or about 186,282 miles per second.

The Sun shining from behind clouds.

No single answer to the question "What is light?" satisfies the many contexts in which light is experienced, explored, and exploited. The physicist is interested in the physical properties of light, the artist in an aesthetic appreciation of the visual world. Through the sense of sight, light is a primary tool for perceiving the world and communicating within it. Light from the Sun warms the Earth, drives global weather patterns, and initiates the life-sustaining process of photosynthesis. On the grandest scale, light's interactions with matter have helped shape the structure of the universe. Indeed, light provides a window on the universe, from cosmological to atomic scales. Almost all of the information about the rest of the universe reaches Earth in the form of electromagnetic radiation. By interpreting that radiation, astronomers can glimpse the earliest epochs of the universe, measure the general expansion of the universe, and determine the chemical composition of stars and the interstellar medium. Just as the invention of the telescope dramatically broadened exploration of the universe, so too the invention of the microscope opened the intricate world of the cell. The analysis of the frequencies of light emitted and absorbed by atoms was a principal impetus for the development of quantum mechanics. Atomic and molecular spectroscopies continue to be primary tools for probing the structure of matter, providing ultrasensitive tests of atomic and molecular models and contributing to studies of fundamental photochemical reactions.

Light transmits spatial and temporal information. This property forms the basis of the fields of optics and optical communications and a myriad of related technologies, both mature and emerging. Technological applications based on the manipulations of light include lasers, holography, and fibre-optic telecommunications systems.

In most everyday circumstances, the properties of light can be derived from the theory of classical electromagnetism, in which light is described as coupled electric and magnetic fields propagating through space as a traveling wave. However, this wave theory, developed in the mid-19th century, is not sufficient to explain the properties of light at very low intensities. At that level a quantum theory is needed to explain the characteristics of light and to explain the interactions of light with atoms and molecules. In its simplest form, quantum theory describes light as consisting of discrete packets of energy, called photons. However, neither a classical wave model nor a classical particle model correctly describes light; light has a dual nature that is revealed only in quantum mechanics. This surprising wave-particle duality is shared by all of the primary constituents of nature (e.g., electrons have both particle-like and wavelike aspects). Since the mid-20th century, a more comprehensive theory of light, known as quantum electrodynamics (QED), has been regarded by physicists as complete. QED combines the ideas of classical electromagnetism, quantum mechanics, and the special theory of relativity.

This topic focuses on the physical characteristics of light and the theoretical models that describe the nature of light. Its major themes include introductions to the fundamentals of geometrical optics, classical electromagnetic waves and the interference effects associated with those waves, and the foundational ideas of the quantum theory of light. More detailed and technical presentations of these topics can be found in the articles optics, electromagnetic radiation, quantum mechanics, and quantum electrodynamics.

Early Particle and Wave Theories

With the dawn of the 17th century, significant progress was reawakened in Europe. Compound microscopes were first constructed in the Netherlands between 1590 and 1608 (probably by Hans and Zacharias Jansen), and most sources credit another Dutchman, Hans Lippershey, with the invention of the telescope in 1608. The Italian astronomer Galileo quickly improved upon the design of the refracting telescope and used it in his discoveries of the moons of Jupiter and the rings of Saturn in 1610. (Refraction refers to the passage of light from one medium into another—in this case, from air into a glass lens.) The German astronomer Johannes Kepler presented an approximate mathematical analysis of the focusing properties of lenses in Dioptrice. An empirical advance was made by the Dutch astronomer Willebrord Snell in 1621 with his discovery of the mathematical relation (Snell's law) between the angles of incidence and transmission for a light ray refracting through an interface between two media. In 1657 the French mathematician Pierre de Fermat presented an intriguing derivation of Snell's law based on his principle of least time, which asserted that light follows the path of minimum time in traveling from one point to another. The posthumous publication of the Jesuit mathematician Francesco Grimaldi's studies in 1665 first described what are now called diffraction effects, in which light passing an obstacle is seen to penetrate into the geometrical shadow. In 1676 the Danish astronomer Ole Rømer used his measurements of the changes in the apparent orbital periods of the moons of Jupiter over the course of a year to deduce an approximate value for the speed of light. The significance of Rømer's work was the realization that the speed of light is not infinite.

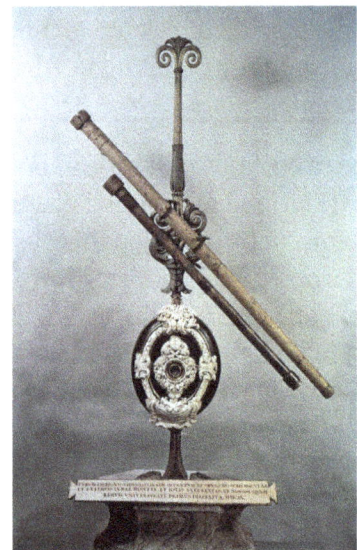

Galileo: telescope Two of Galileo's first telescopes; in the Museo Galileo, Florence.

Seminal physical models of the nature of light were developed in parallel with the many empirical discoveries of the 17th century. Two competing models of light, as a collection of fast-moving particles and as a propagating wave, were advanced. In La Dioptrique, French philosopher-mathematician René Descartes described light as a pressure wave transmitted at infinite speed through a pervasive elastic medium. The prominent English physicist Robert Hooke studied diffraction effects and thin-film interference and concluded in Micrographia that light is a rapid vibration of any medium through which it propagates. In his Traité de la Lumière, the Dutch mathematician-astronomer Christiaan Huygens formulated the first detailed wave theory of light, in the context of which he was also able to derive the laws of reflection and refraction.

The most prominent advocate of a particle theory of light was Isaac Newton. Newton's careful investigations into the properties of light in the 1660s led to his discovery that white light consists of a mixture of colours. He struggled with a formulation of the nature of light, ultimately asserting in Opticks that light consists of a stream of corpuscles, or particles. To reconcile his particle model with the known law of refraction, Newton speculated that transparent objects (such as glass) exert attractive forces on the particles, with the consequence that the speed of light in a transparent medium is always greater than the speed of light in a vacuum. He also postulated that particles of different colours of light have slightly different masses, leading to different speeds in transparent media and hence different angles of refraction. Newton presented his speculations in Opticks in the form of a series of queries rather than as a set of postulates, possibly conveying an ambivalence regarding the ultimate nature of light. Because of his immense authority in the scientific community, there were few challenges to his particle model of light in the century after his death in 1727.

Descartes, René René Descartes.

Newton's corpuscular model survived into the early years of the 19th century, at which time evidence for the wave nature of light became overwhelming. Theoretical and experimental work in the mid to late 19th century convincingly established light as an electromagnetic wave, and the issue seemed to be resolved by 1900. With the arrival of quantum mechanics in the early decades of the 20th century, however, the controversy over the nature of light resurfaced. As will be seen in the following sections, this scientific conflict between particle and wave models of light permeates the history of the subject.

Geometrical Optics: Light as Rays

A detailed understanding of the nature of light was not needed for the development, beginning in the 1600s, of a practical science of optics and optical instrument design. Rather, a set of empirical rules describing the behaviour of light as it traverses transparent materials and reflects off smooth surfaces was adequate to support practical advances in optics. Known collectively today as geometrical optics, the rules constitute an extremely useful, though very approximate, model of light. Their primary applications are the analysis of optical systems—cameras, microscopes, telescopes—and the explanation of simple optical phenomena in nature.

Light Rays

The basic element in geometrical optics is the light ray, a hypothetical construct that indicates the direction of the propagation of light at any point in space. The origin of this concept dates back to early speculations regarding the nature of light. By the 17th century the Pythagorean notion of visual rays had long been abandoned, but the observation that light travels in straight lines led naturally to the development of the ray concept. It is easy to imagine representing a narrow beam of light by a collection of parallel arrows—a bundle of rays. As the beam of light moves from one medium to another, reflects off surfaces, disperses, or comes to a focus, the bundle of rays traces the beam's progress in a simple geometrical manner.

Geometrical optics consists of a set of rules that determine the paths followed by light rays. In any uniform medium the rays travel in straight lines. The light emitted by a small localized source is represented by a collection of rays pointing radially outward from an idealized "point source." A collection of parallel rays is used to represent light flowing with uniform intensity through space; examples include the light from a distant star and the light from a laser. The formation of a sharp shadow when an object is illuminated by a parallel beam of light is easily explained by tracing the paths of the rays that are not blocked by the object.

Reflection and Refraction

Light rays change direction when they reflect off a surface, move from one transparent medium into another, or travel through a medium whose composition is continuously changing. The law of reflection states that, on reflection from a smooth surface, the angle of the reflected ray is equal to the angle of the incident ray. (By convention, all angles in geometrical optics are measured with respect to the normal to the surface—that is, to a line perpendicular to the surface.) The reflected ray is always in the plane defined by the incident ray and the normal to the surface. The law of reflection can be used to understand the images produced by plane and curved mirrors. Unlike mirrors, most natural surfaces are rough on the scale of the wavelength of light, and, as a consequence, parallel incident light rays are reflected in many different directions, or diffusely. Diffuse reflection is responsible for the ability to see most illuminated surfaces from any position—rays reach the eyes after reflecting off every portion of the surface.

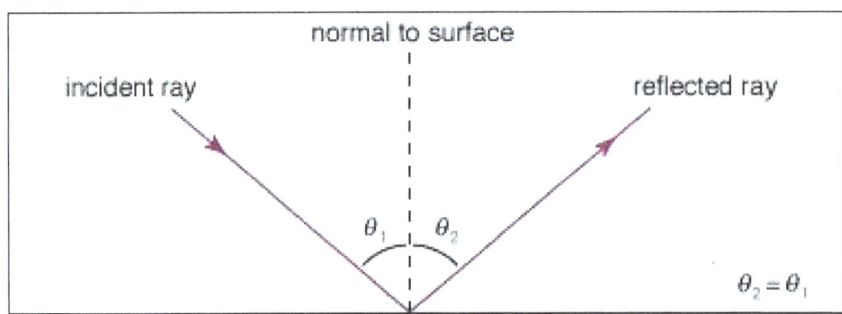

For a smooth surface the angle of incidence (θ_1) equals the angle of reflection (θ_2), as measured with reference to the normal (line perpendicular) to the surface.

When light traveling in one transparent medium encounters a boundary with a second transparent medium (e.g., air and glass), a portion of the light is reflected and a portion is transmitted into the second medium. As the transmitted light moves into the second medium, it changes its direction of travel; that is, it is refracted. The law of refraction, also known as Snell's law, describes the relationship between the angle of incidence (θ_1) and the angle of refraction (θ_2), measured with respect to the normal ("perpendicular line") to the surface, in mathematical terms: $n_1 \sin \theta_1 = n_2 \sin \theta_2$, where n_1 and n_2 are the index of refraction of the first and second media, respectively. The index of refraction for any medium is a dimensionless constant equal to the ratio of the speed of light in a vacuum to its speed in that medium.

By definition, the index of refraction for a vacuum is exactly 1. Because the speed of light in any transparent medium is always less than the speed of light in a vacuum, the indices of refraction of all media are greater than one, with indices for typical transparent materials between one and two. For example, the index of refraction of air at standard conditions is 1.0003, water is 1.33, and glass is about 1.5.

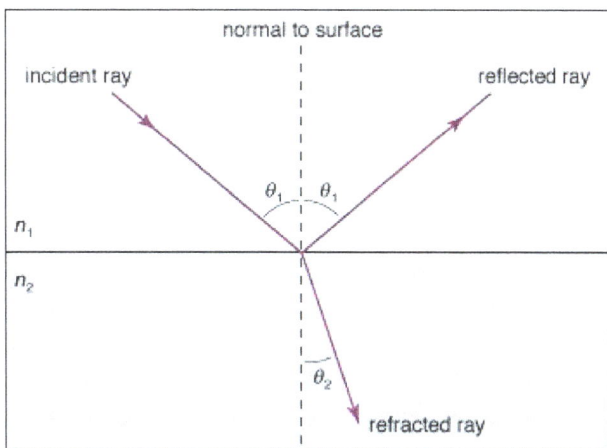

The law of refraction, or Snell's law, predicts the angle at which a light ray
will bend, or refract, as it passes from one medium to another.

The basic features of refraction are easily derived from Snell's law. The amount of bending of a light ray as it crosses a boundary between two media is dictated by the difference in the two indices of refraction. When light passes into a denser medium, the ray is bent toward the normal. Conversely, light emerging obliquely from a denser medium is bent away from the normal. In the special case where the incident beam is perpendicular to the boundary (that is, equal to the normal), there is no change in the direction of the light as it enters the second medium.

Snell's law governs the imaging properties of lenses. Light rays passing through a lens are bent at both surfaces of the lens. With proper design of the curvatures of the surfaces, various focusing effects can be realized. For example, rays initially diverging from a point source of light can be redirected by a lens to converge at a point in space, forming a focused image. The optics of the human eye is centred around the focusing properties of the cornea and the crystalline lens. Light rays from distant objects pass through these two components and are focused into a sharp image on the light-sensitive retina. Other optical imaging systems range from simple single-lens applications, such as the magnifying glass, the eyeglass, and the contact lens, to complex configurations of multiple lenses. It is not unusual for a modern camera to have a half dozen or more separate lens elements, chosen to produce specific magnifications, minimize light losses via unwanted reflections, and minimize image distortion caused by lens aberrations.

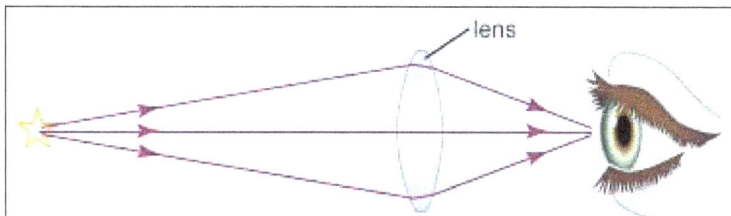

A double convex lens, or converging lens, focuses the diverging, or blurred, light rays from a distant object by refracting (bending) the rays twice. At the front side of the lens, the rays are bent toward the normal (the perpendicular to the surface) because the glass is a denser medium than the air, and, at the back side of the lens, the rays are bent away from the normal as the rays pass into the less-dense medium of the air. This double bending causes the rays to converge at a focal point behind the lens so that a sharper image can be seen or photographed.

Total Internal Reflection

One interesting consequence of the law of refraction is associated with light passing into a medium with a lower index of refraction. As previously mentioned, in this case light rays are bent away from the normal of the interface between the media. At what is called the critical angle of incidence (Θ), the refracted rays make an angle of 90° with the normal—in other words, they just skim the boundary of the two media. The sine of the critical angle is easily derived from the law of refraction: $\sin \Theta = n_2/n_1$.

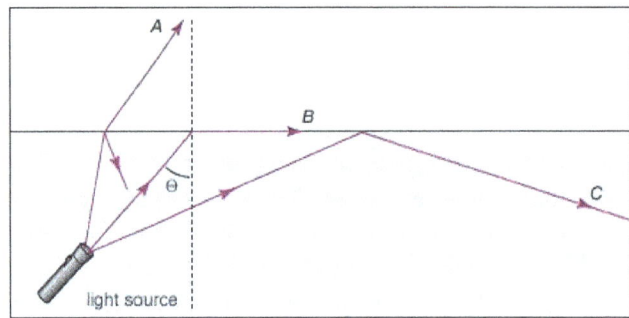

Total internal reflection.

When a light ray strikes the interface between two mediums, it is refracted through an angle that depends on the index of refraction of each material and the ray's angle of incidence, as measured relative to the normal (perpendicular) between the surfaces. At the critical angle of incidence (Θ), light is refracted such that it just remains within the original medium, as shown in ray B. For any light ray, such as C, that strikes the interface at a larger angle, all of the light is reflected internally. For angles less than the critical angle, such as A, some of the light passes through into the second medium.

For any incident angle greater than the critical angle, light rays are completely reflected inside the material. This phenomenon, called total internal reflection, is commonly taken advantage of to "pipe" light in a curved path. When light is directed down a narrow fibre of glass or plastic, the light repeatedly reflects off the fibre-air interface at a large incident angle—larger than the critical angle (for a glass-air interface the critical angle is about 42°). Optical fibres with diameters from 10 to 50 micrometres can transmit light over long distances with little loss of intensity (*see* fibre optics). Optical communications uses sequences of light pulses to transmit information through an optical fibre network. Medical instruments such as endoscopes rely on the total internal reflection of light through an optical fibre bundle to image internal organs.

Dispersion

Through his careful investigation of the refraction of white light as it passed through a glass prism, Newton was famously credited with the discovery that white light consists of a spectrum of colours. The dispersion of white light into its constituent colours is caused by a variation of the index of refraction of glass with colour. This effect, known as chromatic dispersion, results from the fact that the speed of light in glass depends on the wavelength of the light. The speed slightly decreases with decreasing wavelength; this means that the index of refraction, which is inversely proportional to the speed, slightly increases with decreasing wavelength. For glass, the index of refraction for red light (the longest visible wavelength) is about 1 percent less than that for violet light (the shortest visible wavelength).

A prism spreads white light into its various component wavelengths, or colours.

The focusing properties of glass lenses, being determined by their indices of refraction, are slightly dependent on colour. When a single lens images a distant white-light point source, such as a star, the image is slightly distorted because of dispersion in the lens; this effect is called chromatic aberration. In an effort to improve upon the chromatic aberration of the refracting telescope, Isaac Newtoninvented the reflecting telescope, in which the imaging and magnification are accomplished with mirrors.

Dispersion is not restricted to glass; all transparent media exhibit some dispersion. Many beautiful optical effects are explained by the phenomena of dispersion, refraction, and reflection. Principal among them is the rainbow, for which René Descartes and Newton are credited with the first solid quantitative analyses. A rainbow is formed when sunlight is refracted by spherical water droplets in the atmosphere; two refractions and one reflection, combined with the chromatic dispersion of water, produce the primary arcs of colour. The laws of geometrical optics also explain the formation of mirages and halos and the rarely observed "green flash" of a setting Sun.

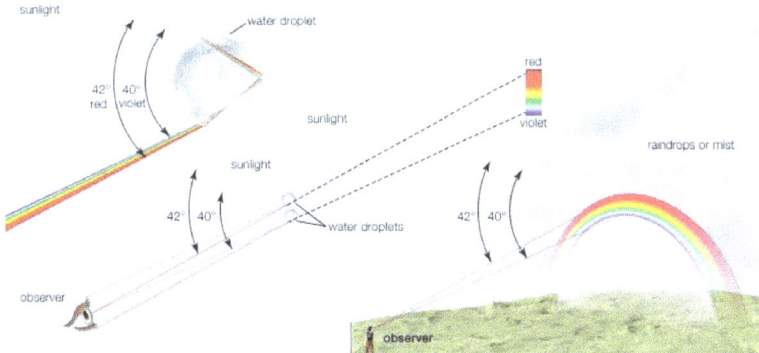

Rainbow effect As sunlight enters water droplets in the atmosphere, its constituent colours (wavelengths) are bent (refracted) by slightly different amounts during its passage from the air into the water. A portion of the light striking the back of each water droplet is internally reflected and then refracted a second time as it reemerges into the air. Violet light is refracted most and reemerges at an angle of about 40° compared with the incident sunlight; red light is refracted least and reemerges at an angle of 42° compared with the incident sunlight. In order for observers to see a rainbow, they must have the Sun behind them such that the angle between the incident sunlight and their line of sight is about 42°. Thus, a rainbow forms a full circular arc with a central angle from the observer of 42°. However, the lower portion of the arc is obscured by the surface of the Earth; consequently, the maximum arc (a semicircle) can be seen near sunset.

The rules of geometrical optics, developed through centuries of observation, can be derived from the classical electromagnetic-wave model of light. However, as long as the physical dimensions of the objects that light encounters (and the apertures through which it passes) are significantly greater than the wavelength of the electromagnetic wave, there is no need for the mathematical formalism of the wave model. In those circumstances, light is adequately modeled as a collection of rays following the rules of geometrical optics. Most everyday optical phenomena can be handled within this approximation, since the wavelengths of visible light are relatively short (400 to 700 nm). However, as the dimensions of objects and apertures approach the wavelength of light, the wave character of light cannot be disregarded. Many optical effects, often subtle in nature, cannot be understood without a wave model. For example, on close inspection the shadows of objects in parallel light are seen not to be infinitely sharp. This is a consequence of the "bending" of waves around corners—a phenomenon best explained by the wave model. Another class of phenomena involves the polarization of light waves. These issues are addressed below.

Light as a Wave

Isaac Newton's corpuscular model of light was championed by most of the European scientific community throughout the 1700s, but by the start of the 19th century it was facing challenges. About 1802 Thomas Young, an English physician and physicist, showed that an interference pattern is produced when light from two sources overlaps. Though it took some time for Young's contemporaries fully to accept the implications of his landmark discovery, it conclusively demonstrated that light has wavelike characteristics. Young's work ushered in a period of intense experimental and theoretical activity that culminated 60 years later in a fully developed wave theory of light. By the latter years of the 19th century, corpuscular theories were abandoned. Before describing Young's work, an introduction to the relevant features of waves is in order.

Characteristics of Waves

From ripples on a pond to deep ocean swells, sound waves, and light, all waves share some basic characteristics. Broadly speaking, a wave is a disturbance that propagates through space. Most waves move through a supporting medium, with the disturbance being a physical displacement of the medium. The time dependence of the displacement at any single point in space is often an oscillation about some equilibrium position. For example, a sound wave travels through the medium of air, and the disturbance is a small collective displacement of air molecules—individual molecules oscillate back and forth as the wave passes.

Unlike particles, which have well-defined positions and trajectories, waves are not localized in space. Rather, waves fill regions of space, and their evolutions in time are not described by simple trajectories. Nevertheless, some waves are more localized than others, and so it is useful to distinguish two broad classes. (1) A wave pulse is a relatively localized disturbance. For example, when a stone is dropped into a pond, the resulting ripples, which constitute a surface wave, extend over only a small portion of the surface at any instant of time. (2) At the opposite extreme, periodic waves can extend over great distances. In the example above, if the water surface is repeatedly disturbed at one point for a long period of time, the surface ripples eventually will blanket a large area.

A simple and useful example of a periodic wave is a harmonic wave. The wavelength λ of the wave is the physical separation between successive crests. The maximum displacement of the wave, or amplitude, is denoted by A. The time between successive oscillations is called the period τ of the wave. The number of oscillations per second is the wave frequency f, which is the reciprocal of the period, 1/τ.

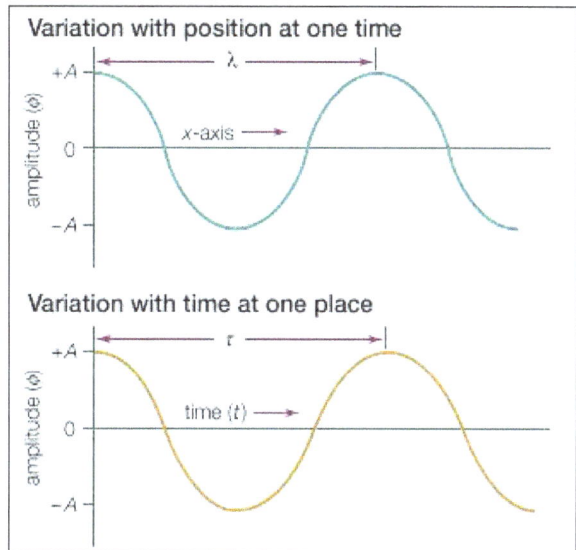

Snapshots of a harmonic wave can be taken at a fixed time to display the wave's variation with position (top) or at a fixed location to display the wave's variation with time (bottom).

Harmonic waves propagate with well-defined velocities that are related to their frequency and wavelength. Fixing attention on a single point in space, the number of wave crests that pass that point per second is the wave frequency f. The distance traveled past that point by any one crest in one second—the wave velocity v—is equal to the distance between crests λ multiplied by the frequency: v = λf.

The properties of harmonic waves are illustrated in the mathematical expression for the displacement in both space and time. For a harmonic wave traveling in the x-direction, the spatial and time dependence of the displacement is:

$$\phi(x,t) = A\cos\left(\frac{2\pi x}{\lambda} - 2\pi ft\right).$$

Interference

A defining characteristic of all waves is superposition, which describes the behaviour of overlapping waves. The superposition principle states that when two or more waves overlap in space, the resultant disturbance is equal to the algebraic sum of the individual disturbances. This simple underlying behaviour leads to a number of effects that are collectively called interference phenomena.

There are two extreme limits to interference effects. In constructive interference the crests of two waves coincide, and the waves are said to be in phase with each other. Their superposition results in a reinforcement of the disturbance; the amplitude of the resulting combined wave is the sum of the individual amplitudes. Conversely, in destructive interference the crest of one wave coincides with the valley of a second wave, and they are said to be out of phase. The amplitude of the combined wave equals the difference between the amplitudes of the individual waves. In the special case where those individual amplitudes are equal, the destructive interference is complete, and the net disturbance to the medium is zero.

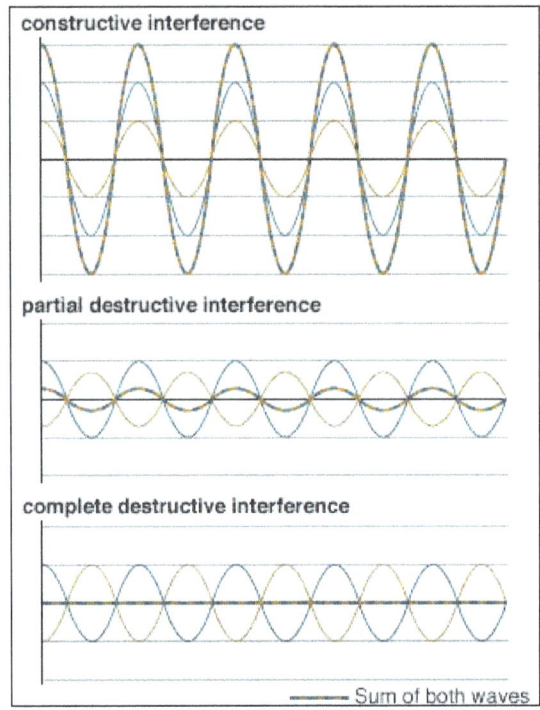

When two waves of identical wavelength are in phase, they form a new wave with an amplitude equal to the sum of their individual amplitudes (constructive interference). When two waves are of completely opposite phase, they either form a new wave of reduced amplitude (partial destructive interference) or cancel each other out (complete destructive interference). Much more complicated constructive and destructive interference patterns emerge when waves with different wavelengths interact.

Young's Double-slit Experiment

The observation of interference effects definitively indicates the presence of overlapping waves. Thomas Young postulated that light is a wave and is subject to the superposition principle; his great experimental achievement was to demonstrate the constructive and destructive interference of light. In a modern version of Young's experiment, differing in its essentials only in the source of light, a laser equally illuminates two parallel slits in an otherwise opaque surface. The light passing through the two slits is observed on a distant screen. When the widths of the slits are significantly greater than the wavelength of the light, the rules of geometrical optics hold—the light casts two shadows, and there are two illuminated regions on the screen. However, as the slits are narrowed in width, the light diffracts into the geometrical shadow, and the light waves overlap on the screen.

The superposition principle determines the resulting intensity pattern on the illuminated screen. Constructive interference occurs whenever the difference in paths from the two slits to a point on the screen equals an integral number of wavelengths (0, λ, 2λ,...). This path difference guarantees that crests from the two waves arrive simultaneously. Destructive interference arises from path differences that equal a half-integral number of wavelengths ($\lambda/2$, $3\lambda/2$,...). Young used geometrical arguments to show that the superposition of the two waves results in a series of equally spaced bands, or fringes, of high intensity, corresponding to regions of constructive interference, separated by dark regions of complete destructive interference.

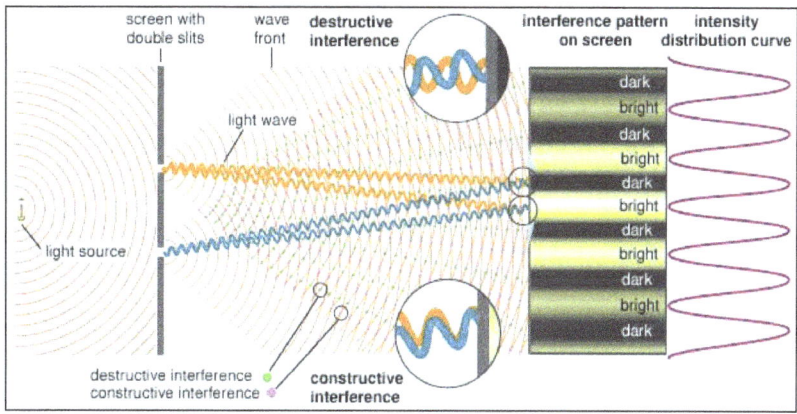

Young's double-slit experiment When monochromatic light passing through two narrow slits illuminates a distant screen, a characteristic pattern of bright and dark fringes is observed. This interference pattern is caused by the superposition of overlapping light waves originating from the two slits. Regions of constructive interference, corresponding to bright fringes, are produced when the path difference from the two slits to the fringe is an integral number of wavelengths of the light. Destructive interference and dark fringes are produced when the path difference is a half-integral number of wavelengths.

An important parameter in the double-slit geometry is the ratio of the wavelength of the light λ to the spacing of the slits d. If λ/d is much smaller than 1, the spacing between consecutive interference fringes will be small, and the interference effects may not be observable. Using narrowly separated slits, Young was able to separate the interference fringes. In this way he determined the wavelengths of the colours of visible light. The very short wavelengths of visible light explain why interference effects are observed only in special circumstances—the spacing between the sources of the interfering light waves must be very small to separate regions of constructive and destructive interference.

Observing interference effects is challenging because of two other difficulties. Most light sources emit a continuous range of wavelengths, which result in many overlapping interference patterns, each with a different fringe spacing. The multiple interference patterns wash out the most pronounced interference effects, such as the regions of complete darkness. Second, for an interference pattern to be observable over any extended period of time, the two sources of light must be coherent with respect to each other. This means that the light sources must maintain a constant phase relationship. For example, two harmonic waves of the same frequency always have a fixed phase relationship at every point in space, being either in phase, out of phase, or in some intermediate relationship. However, most light sources do not emit true harmonic waves; instead, they emit waves that undergo random phase changes millions of times per second. Such light is called incoherent. Interference still occurs when light waves from two incoherent sources overlap in space, but the interference pattern fluctuates randomly as the phases of the waves shift randomly. Detectors of light, including the eye, cannot register the quickly shifting interference patterns, and only a time-averaged intensity is observed. Laser light is approximately monochromatic (consisting of a single wavelength) and is highly coherent; it is thus an ideal source for revealing interference effects.

After 1802, Young's measurements of the wavelengths of visible light could be combined with the relatively crude determinations of the speed of light available at the time in order to calculate the approximate frequencies of light. For example, the frequency of green light is about 6×10^{14} Hz (hertz, or cycles per second). This frequency is many orders of magnitude larger than the frequencies of common mechanical waves. For comparison, humans can hear sound waves with frequencies up to about 2×10^4 Hz. Exactly what was oscillating at such a high rate remained a mystery for another 60 years.

Thin-film Interference

Observable interference effects are not limited to the double-slit geometry used by Thomas Young. The phenomenon of thin-film interference results whenever light reflects off two surfaces separated by a distance comparable to its wavelength. The "film" between the surfaces can be a vacuum, air, or any transparent liquid or solid. In visible light, noticeable interference effects are restricted to films with thicknesses on the order of a few micrometres. A familiar example is the film of a soap bubble. Light reflected from a bubble is a superposition of two waves—one reflecting off the front surface and a second reflecting off the back surface. The two reflected waves overlap in space and interfere. Depending on the thickness of the soap film, the two waves may interfere constructively or destructively. A full analysis shows that, for light of a single wavelength λ, there are constructive interference for film thicknesses equal to $\lambda/4$, $3\lambda/4$, $5\lambda/4,...$ and destructive interference for thicknesses equal to $\lambda/2$, λ, $3\lambda/2,....$

When white light illuminates a soap film, bright bands of colour are observed as different wavelengths suffer destructive interference and are removed from the reflection. The remaining reflected light appears as the complementary colour of the removed wavelength (e.g., if red light is removed by destructive interference, the reflected light will appear as cyan). Thin films of oil on water produce a similar effect. In nature, the feathers of certain birds, including peacocks and hummingbirds, and the shells of some beetles display iridescence, in which the colour on reflection changes with the viewing angle. This is caused by the interference of reflected light waves from thinly layered structures or regular arrays of reflecting rods. In a similar fashion, pearls and abalone shells are iridescent from the interference caused by reflections from multiple layers of nacre. Gemstones such as opal exhibit beautiful interference effects arising from the scattering of light from regular patterns of microscopic spherical particles.

There are many technological applications of interference effects in light. Common antireflection coatings on camera lenses are thin films with thicknesses and indices of refraction chosen to produce destructive interference on reflection for visible light. More-specialized coatings, consisting of multiple layers of thin films, are designed to transmit light only within a narrow range of wavelengths and thus act as wavelength filters. Multilayer coatings are also used to enhance the reflectivity of mirrors in astronomical telescopes and in the optical cavities of lasers. The precision techniques of interferometry measure small changes in relative distances by monitoring the fringe shifts in the interference patterns of reflected light. For example, the curvatures of surfaces in optical components are monitored to fractions of an optical wavelength with interferometric methods.

Diffraction

The subtle pattern of light and dark fringes seen in the geometrical shadow when light passes an obstacle, first observed by the Jesuit mathematician Francesco Grimaldi in the 17th century, is an example of the wave phenomenon of diffraction. Diffraction is a product of the superposition of waves—it is an interference effect. Whenever a wave is obstructed, those portions of the wave not affected by the obstruction interfere with one another in the region of space beyond the obstruction. The mathematics of diffraction is considerably complicated, and a detailed, systematic theory was not worked out until 1818 by the French physicist Augustin-Jean Fresnel.

The Dutch scientist Christiaan Huygens first stated the fundamental principle for understanding diffraction: every point on a wave front can be considered a secondary source of spherical wavelets. The shape of the advancing wave front is determined by the envelope of the overlapping spherical wavelets. If the wave is unobstructed, Huygens's principle will not be needed for determining its evolution—the rules of geometrical optics will suffice. (However, note that the light rays of geometrical optics are always perpendicular to the advancing wavefront; in this sense, the progress of a light ray is ultimately always determined by Huygens's principle.) Huygens's principle becomes necessary when a wave meets an obstacle or an aperture in an otherwise opaque surface. Thus, for a plane wave passing through a small aperture, only wavelets originating within the aperture contribute to the transmitted wave, which is seen to spread into the region of the aperture's geometric shadow.

Fresnel incorporated Young's principle of interference into Huygens's construction and calculated the detailed intensity patterns produced by interfering secondary wavelets. For a viewing screen a distance L from a slit of width a, light of wavelength λ produces a central intensity maximum that is approximately λL/a in width. This result highlights the most important qualitative feature of diffraction: the effect is normally apparent only when the sizes of obstacles or apertures are on the order of the wavelength of the wave. For example, audible sound waves have wavelengths of about one metre, which easily diffract around commonplace objects. This is why sound is heard around corners. On the other hand, visible light has wavelengths of a fraction of a micrometre, and it therefore does not noticeably bend around large objects. Only the most careful measurements by Young, Fresnel, and their early 19th-century contemporaries revealed the details of the diffraction of visible light.

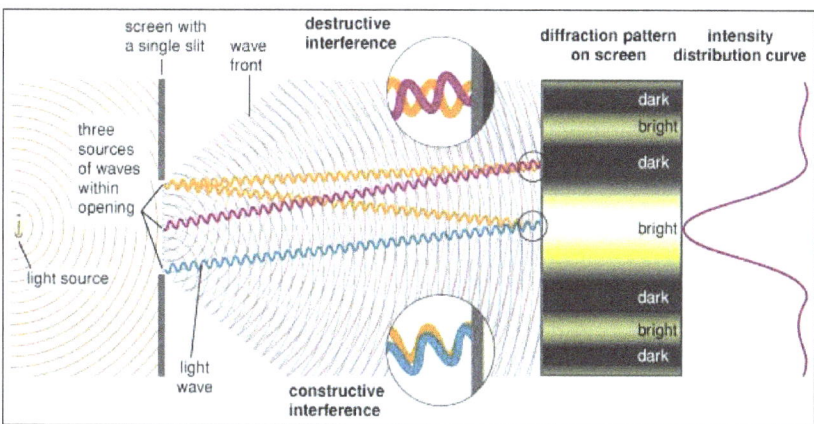

Single-slit diffraction When monochromatic light passing through a single slit illuminates a screen, a characteristic diffraction pattern is observed. Diffraction is a product of the superposition of waves—i.e., it is an interference effect. The detailed pattern of constructive and destructive inter-ference fringes can be derived by treating every point on the wave front passing through the slit as a secondary source of spherical waves. The paths from three representative secondary sources to the viewing screen are shown here. The central bright fringe in a single-slit diffraction pattern is produced by the constructive interference of all of the secondary sources. The width of the central fringe is inversely proportional to the width of the slit. Diffraction effects become pronounced only when the width of the slit is an appreciable fraction of the wavelength of the light.

Diffraction Effects

Poisson's Spot

Fresnel presented much of his work on diffraction as an entry to a competition on the subject spon-sored by the French Academy of Sciences. The committee of judges included a number of prom-inent advocates of Newton's corpuscular model of light, one of whom, Siméon-Denis Poisson, pointed out that Fresnel's model predicted a seemingly absurd result: If a parallel beam of light falls on a small spherical obstacle, there will be a bright spot at the centre of the circular shadow—a spot nearly as bright as if the obstacle were not there at all. An experiment was subsequently per-formed by the French physicist François Arago, and Poisson's spot was seen, vindicating Fresnel.

Circular Apertures and Image Resolution

Circular apertures also produce diffraction patterns. When a parallel beam of light passes through a converging lens, the rules of geometrical optics predict that the light comes to a tight focus be-hind the lens, forming a point image. In reality, the pattern in the lens's image plane is compli-cated by diffraction effects. The lens, considered as a circular aperture with diameter D, produces a two-dimensional diffraction pattern with a central intensity maximum of angular width about λ/D. Angular width refers to the angle, measured in radians, that is defined by the two intensity minima on either side of the central maximum.

Diffraction effects from circular apertures have an important practical consequence: the intensity patterns in optical images produced by circular lenses and mirrors are limited in their ability to resolve closely spaced features. Each point in the object is imaged into a diffraction pattern of finite width, and the final image is a sum of individual diffraction patterns. Baron Rayleigh, a leading figure of late 19th-century physics, showed that the images of two point sources are resolvable only if their angular separation, relative to an imaging element of diameter D, is greater than about $1.2\lambda/D$ ("Rayleigh's criterion").

Circular aperture diffraction effects limit the resolving power of telescopes and microscopes. This is one of the reasons why the best astronomical telescopes have large-diameter mirrors; in addi-tion to the obvious advantage of an increased light-gathering capability, larger mirrors decrease the resolvable angular separation of astronomical objects. To minimize diffraction effects, optical microscopes are sometimes designed to use ultraviolet light rather than longer-wavelength visible light. Nevertheless, diffraction is often the limiting factor in the ability of a microscope to resolve the fine details of objects.

The late 19th-century French painter Georges Seurat created a new technique, known as pointillism, based on diffraction effects. His paintings consist of thousands of closely spaced small dots of colour. When viewed up close, the individual points of colour are apparent to the eye. Viewed from afar, the individual points cannot be resolved because of the diffraction of the images produced by the lens of the eye. The overlapping images on the retina combine to produce colours other than those used in the individual dots of paint. The same physics underlies the use of closely spaced arrays of red, blue, and green phosphors on television screens and computer monitors; diffraction effects in the eye mix the three primary colours to produce a wide range of hues.

Atmospheric Diffraction Effects

diffraction rings Diffraction rings, called a glory, occur most commonly when the Sun shines on a cloud or fog. The radius of a ring is dependent on the size of the cloud droplets—the smaller the droplets, the larger the radius. Moreover, the droplets must be nearly uniform in size for the phenomenon to appear.

Diffraction is also responsible for certain optical effects in Earth's atmosphere. A set of concentric coloured rings, known as an atmospheric corona, often overlapping to produce a single diffuse whitish ring, is sometimes observed around the Moon. The corona is produced as light reflected from the Moon diffracts through water droplets or ice crystals in Earth's upper atmosphere. When the droplets are of uniform diameter, the different colours are clearly distinct in the diffraction pattern. A related and beautiful atmospheric phenomenon is the glory. Seen in backscattered light from water droplets, commonly forming a fog or mist, the glory is a set of rings of coloured light surrounding the shadow of the observer. The rings of light, with angular diameters of a few degrees, are created by the interplay of refraction, reflection, and diffraction in the water droplets. The glory, once a phenomenon rarely observed, is now frequently seen by airline travelers as coloured rings surrounding their airplane's shadow on a nearby cloud. Finally, as pointed out in the section Dispersion, the primary and secondary arcs of a rainbow are adequately explained by geometrical optics. However, the more subtle supernumerary bows—weak arcs of light occasionally seen below the primary arc of colours—are caused by diffraction effects in the water droplets that form the rainbow.

Doppler Effect

In 1842 Austrian physicist Christian Doppler established that the apparent frequency of sound waves from an approaching source is greater than the frequency emitted by the source and that the apparent frequency of a receding source is lower. The Doppler effect, which is easily noticed with approaching or receding police sirens, also applies to light waves. The light from an approaching source is shifted up in frequency, or blueshifted, while light from a receding source is shifted down in frequency, or redshifted. The frequency shift depends on the velocity of the source relative to the observer; for velocities much less than the speed of light, the shift is proportional to the velocity.

The observation of Doppler shifts in atomic spectral lines is a powerful tool to measure relative motion in astronomy. Most notably, redshifted light from distant galaxies is the primary evidence for the general expansion of the universe. There are a host of other astronomical applications, including the determination of binary star orbits and the rotation rates of galaxies. The most common terrestrial application of the Doppler effect occurs in radar systems. Electromagnetic waves reflected from a moving object undergo Doppler shifts that can then be used to determine the object's speed. In these applications, ranging from monitoring automobile speeds to monitoring wind speeds in the atmosphere, radio waves or microwaves are used instead of visible light.

Light as Electromagnetic Radiation

In spite of theoretical and experimental advances in the first half of the 19th century that established the wave properties of light, the nature of light was not yet revealed—the identity of the wave oscillations remained a mystery. This situation dramatically changed in the 1860s when the Scottish physicist James Clerk Maxwell, in a watershed theoretical treatment, unified the fields of electricity, magnetism, and optics. In his formulation of electromagnetism, Maxwell described light as a propagating wave of electric and magnetic fields. More generally, he predicted the existence of electromagnetic radiation: coupled electric and magnetic fields traveling as waves at a speed equal to the known speed of light. In 1888 German physicist Heinrich Hertz succeeded in demonstrating the existence of long-wavelength electromagnetic waves and showed that their properties are consistent with those of the shorter-wavelength visible light.

Electric and Magnetic Fields

The subjects of electricity and magnetism were well developed by the time Maxwell began his synthesizing work. English physician William Gilbert initiated the careful study of magnetic phenomena in the late 16th century. In the late 1700s an understanding of electric phenomena was pioneered by Benjamin Franklin, Charles-Augustin de Coulomb, and others. Siméon-Denis Poisson, Pierre-Simon Laplace, and Carl Friedrich Gauss developed powerful mathematical descriptions of electrostatics and magnetostatics that stand to the present time. The first connection between electric and magnetic effects was discovered by Danish physicist Hans Christian Ørsted in 1820 when he found that electric currents produce magnetic forces. Soon after, French physicist André-Marie Ampère developed a mathematical formulation (Ampère's law) relating currents to magnetic effects. In 1831 the great English experimentalist Michael Faraday discovered electromagnetic induction, in which a moving magnet (more generally, a changing magnetic flux) induces an electric current in a conducting circuit.

Faraday's conception of electric and magnetic effects laid the groundwork for Maxwell's equations. Faraday visualized electric charges as producing fields that extend through space and transmit electric and magnetic forces to other distant charges. The notion of electric and magnetic fields is central to the theory of electromagnetism, and so it requires some explanation. A field is used to represent any physical quantity whose value changes from one point in space to another. For example, the temperature of Earth's atmosphere has a definite value at every point above the surface of Earth; to specify the atmospheric temperature completely thus requires specifying a distribution of numbers—one for each spatial point. The temperature "field" is simply a mathematical accounting of those numbers; it may be expressed as a function of the spatial coordinates. The values of the temperature field can also vary with time; therefore, the field is more generally expressed as a function of spatial coordinates and time: T(x, y, z, t), where T is the temperature field, x, y, and z are the spatial coordinates, and t is the time.

Temperature is an example of a scalar field; its complete specification requires only one number for each spatial point. Vector fields, on the other hand, describe physical quantities that have a direction and magnitude at each point in space. A familiar example is the velocity field of a fluid. Electric and magnetic fields are also vector fields; the electric field is written as E(x, y, z, t) and the magnetic field as B(x, y, z, t).

Maxwell's Equations

In the early 1860s, Maxwell completed a study of electric and magnetic phenomena. He presented a mathematical formulation in which the values of the electric and magnetic fields at all points in space can be calculated from a knowledge of the sources of the fields. By Faraday's time, it was known that electric charges are the source of electric fields and that electric currents (charges in motion) are the source of magnetic fields. Faraday's electromagnetic induction showed that there is a second source of electric fields—changing magnetic fields. In a significant step in the development of his theory, Maxwell postulated that changing electric fields are sources of magnetic fields. In its modern form, Maxwell's electromagnetic theory is expressed as four partial differential equations for the fields E and B. Known as Maxwell's equations, these four statements relating the fields to their sources, along with the expression for the forces exerted by the fields on electric charges, constitute the whole of classical electromagnetism.

Electromagnetic Waves and the Electromagnetic Spectrum

Electromagnetic Waves

A manipulation of the four equations for the electric and magnetic fields led Maxwell to wave equations for the fields, the solutions of which are traveling harmonic waves. Though the mathematical treatment is detailed, the underlying origin of the waves can be understood qualitatively: changing magnetic fields produce electric fields, and changing electric fields produce magnetic fields. This implies the possibility of an electromagnetic field in which a changing electric field continually gives rise to a changing magnetic field, and vice versa.

Electromagnetic waves do not represent physical displacements that propagate through a medium like mechanical sound and water waves; instead, they describe propagating oscillations in the strengths of electric and magnetic fields. Maxwell's wave equation showed that the speed of the waves, labeled c, is determined by a combination of constants in the laws of electrostatics and magnetostatics—in modern notation:

$$c = \frac{1}{\sqrt{\varepsilon_0 \mu_0}}$$

where ε_0, the permittivity of free space, has an experimentally determined value of 8.85×10^{-12}

square coulomb per newton square metre, and μ0, the magnetic permeability of free space, has a value of 1.26 × 10−6 newton square seconds per square coulomb. The calculated speed, about 3 × 108 metres per second, agreed with the known speed of light. In an 1864 lecture before the Royal Society of London, "A Dynamical Theory of the Electro-Magnetic Field," Maxwell asserted:

> We have strong reason to conclude that light itself—including radiant heat and other radiation, if any—is an electromagnetic disturbance in the form of waves propagated through the electro-magnetic field according to electro-magnetic laws.

Maxwell's achievement ranks as one of the greatest advances of physics. For the physicist of the late 19th century, the study of light became a study of an electromagnetic phenomenon—the fields of electricity, magnetism, and optics were unified in one grand design. While an understanding of light has undergone some profound changes since the 1860s as a result of the discovery of light's quantum mechanical nature, Maxwell's electromagnetic wave model remains completely adequate for many purposes.

The Electromagnetic Spectrum

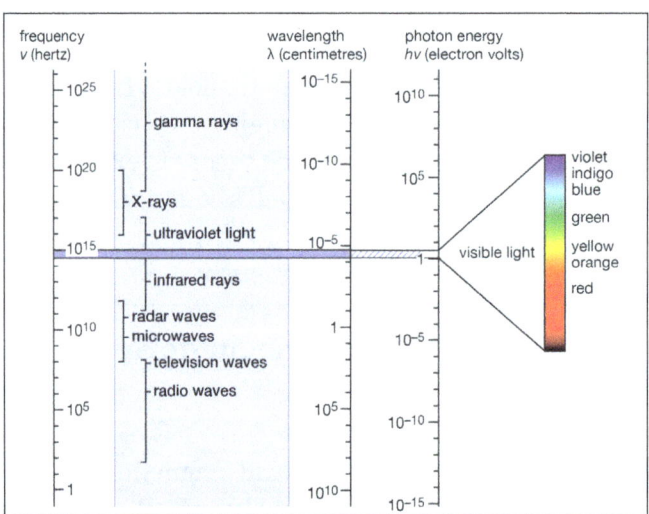

The position of light in the electromagnetic spectrum. The narrow range of visible light is shown enlarged at the right.

Heinrich Hertz's production in 1888 of what are now called radio waves, his verification that these waves travel at the same speed as visible light, and his measurements of their reflection, refraction, diffraction, and polarization properties were a convincing demonstration of the existence of Maxwell's waves. Visible light is but one example of a much broader set of phenomena—an electromagnetic spectrum with no theoretical upper or lower limit to frequencies and wavelengths. While there are no theoretical distinctions between electromagnetic waves of any wavelength, the spectrum is conventionally divided into different regions on the basis of historical developments, the methods of production and detection of the waves, and their technological uses.

Sources of Electromagnetic Waves

The sources of classical electromagnetic waves are accelerating electric charges. (Note that acceleration refers to a change in velocity, which occurs whenever a particle's speed or its direction of motion changes.) A common example is the generation of radio waves by oscillating electric charges in an antenna. When a charge moves in a linear antenna with an oscillation frequency f, the oscillatory motion constitutes an acceleration, and an electromagnetic wave with the same frequency propagates away from the antenna. At frequencies above the microwave region, with a few prominent exceptions, the classical picture of an accelerating electric charge producing an electromagnetic wave is less and less applicable. In the infrared, visible, and ultraviolet regions, the primary radiators are the charged particles in atoms and molecules. In this regime a quantum mechanical radiation model is far more relevant.

The Speed of Light

Early Measurements

Measurements of the speed of light have challenged scientists for centuries. The assumption that the speed is infinite was dispelled by the Danish astronomer Ole Rømer in 1676. French physicist Armand-Hippolyte-Louis Fizeau was the first to succeed in a terrestrial measurement in 1849, sending a light beam along a 17.3-km round-trip path across the outskirts of Paris. At the light source, the exiting beam was chopped by a rotating toothed wheel; the measured rotational rate of the wheel at which the beam, upon its return, was eclipsed by the toothed rim was used to determine the beam's travel time. Fizeau reported a light speed that differs by only about 5 percent from the currently accepted value. One year later, French physicist Léon Foucault improved the accuracy of the technique to about 1 percent.

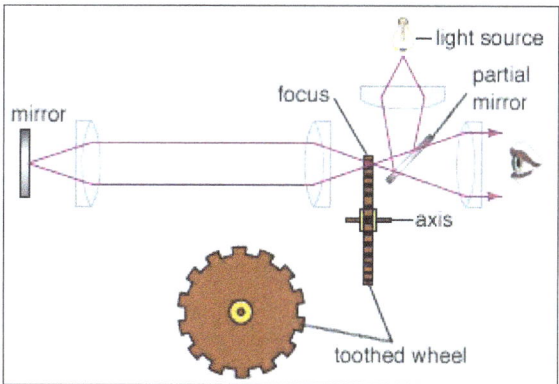

In 1849 Armand Fizeau sent light pulses through a rotating toothed wheel. A distant mirror on the other side reflected the pulses back through gaps in the wheel. By rotating the wheel at a certain speed, each light pulse that went through a gap on the way out was blocked by the next tooth as it came around. Knowing the distance to the mirror and the speed of rotation of the wheel enabled Fizeau to obtain one of the earliest measurements of the speed of light.

In the same year, Foucault showed that the speed of light in water is less than its speed in air by the ratio of the indices of refraction of air and water:

$$v_{water} = \frac{air}{water} v_{air} \approx 0.75 v_{air}.$$

This measurement established the index of refraction of a material as the ratio of the speed of light in vacuum to the speed within the material. The more general finding, that light is slowed in transparent media, directly contradicted Isaac Newton's assertion that light corpuscles travel faster in media than in vacuum and settled any lingering 19th-century doubts about the corpuscle–wave debate.

The Michelson-Morley Experiment

The German-born American physicist A.A. Michelson set the early standard for measurements of the speed of light in the late 1870s, determining a speed within 0.02 percent of the modern value. Michelson's most noteworthy measurements of the speed of light, however, were yet to come. From the first speculations on the wave nature of light by Huygens through the progressively more refined theories of Young, Fresnel, and Maxwell, it was assumed that an underlying physical medium supports the transmission of light, in much the same way that air supports the transmission of sound. Called the ether, or the luminiferous ether, this medium was thought to permeate all of space. The inferred physical properties of the ether were problematic—to support the high-frequency transverse oscillations of light, it would have to be very rigid, but its lack of effect on planetary motion and the fact that it was not observed in any terrestrial circumstances required it to be tenuous and chemically undetectable. In 1887 Michelson, in collaboration with American chemist Edward Morley, completed a precise set of optical measurements designed to detect the motion of Earth through the ether as it orbited the Sun.

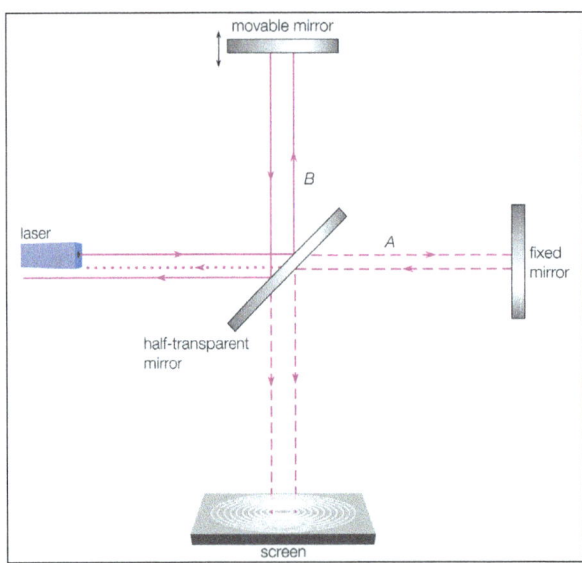

The Michelson interferometer consists of a half-transparent mirror oriented at a 45° angle to a light beam so that the light is divided into two equal parts (A and B), one of which is transmitted to a fixed mirror and the other of which is reflected to a movable mirror. The half-transparent mirror has the same effect on the returning beams, splitting each of them into two beams. Thus, two diminished light beams reach the screen, where interference patterns can be observed by varying the position of the movable mirror.

The measurements in the Michelson-Morley experiment were based on the assumption that an observer at rest in the ether would determine a different speed from an observer moving through the ether. Because Earth's speed relative to the Sun is about 29,000 metres per second, or about 0.01 percent of the speed of light, Earth provides a convenient vantage point for measuring any change in the relative speed of light due to motion. Using a Michelson optical interferometer, interference effects between two light beams traveling parallel to, and perpendicular to, Earth's orbital motion were monitored during the course of its orbit. The instrument was capable of detecting a difference in light speeds along the two paths of the interferometer as small as 5,000 metres per second (less than 2 parts in 100,000 of the speed of light). No difference was found. If Earth indeed moved through the ether, that motion seemed to have no effect on the measured speed of light.

What is now known as the most famous experimental null result in physics was reconciled in 1905 when Albert Einstein, in his formulation of special relativity, postulated that the speed of light is the same in all reference frames; i.e., the measured speed of light is independent of the relative motion of the observer and the light source. The hypothetical ether, with its preferred reference frame, was eventually abandoned as an unnecessary construct.

Fundamental Constant of Nature

Since Einstein's work, the speed of light is considered a fundamental constant of nature. Its significance is far broader than its role in describing a property of electromagnetic waves. It serves as the single limiting velocity in the universe, being an upper bound to the propagation speed of signals and to the speeds of all material particles. In the famous relativity equation, $E = mc^2$, the speed of light (c) serves as a constant of proportionality linking the formerly disparate concepts of mass (m) and energy (E).

Measurements of the speed of light were successively refined in the 20th century, eventually reaching a precision limited by the definitions of the units of length and time—the metre and the second. In 1983 the 17th General Conference on Weights and Measures fixed the speed of light as a defined constant at exactly 299,792,458 metres per second. The metre became a derived unit, equaling the distance traveled by light in 1/299,792,458 of a second.

Polarization

Transverse Waves

Waves come in two varieties. In a longitudinal wave the oscillating disturbance is parallel to the direction of propagation. A familiar example is a sound wave in air—the oscillating motions of the air molecules are induced in the direction of the advancing wave. Transverse waves consist of disturbances that are at right angles to the direction of propagation; for example, as a wave travels horizontally through a body of water, its surface bobs up and down.

A number of puzzling optical effects, first observed in the mid-17th century, were resolved when light was understood as a wave phenomenon and the directions of its oscillations were uncovered. The first so-called polarization effect was discovered by the Danish physician Erasmus Bartholin in 1669. Bartholin observed double refraction, or birefringence, in calcite (a common crystalline form of calcium carbonate). When light passes through calcite, the crystal splits the light, producing two images offset from each other. Newton was aware of this effect and speculated that perhaps his corpuscles of light had an asymmetry or "sidedness" that could explain the formation of the two images. Huygens, a contemporary of Newton, could account for double refraction with his elementary wave theory, but he did not recognize the true implications of the effect. Double refraction remained a mystery until Thomas Young, and independently the French physicist Augustin-Jean

Fresnel, suggested that light waves are transverse. This simple notion provided a natural and un-complicated framework for the analysis of polarization effects. (The polarization of the entering light wave can be described as a combination of two perpendicular polarizations, each with its own wave speed. Because of their different wave speeds, the two polarization components have different indices of refraction, and they therefore refract differently through the material, produc-ing two images.) Fresnel quickly developed a comprehensive model of transverse light waves that accounted for double refraction and a host of other optical effects. Forty years later, Maxwell's electromagnetic theory elegantly provided the basis for the transverse nature of light.

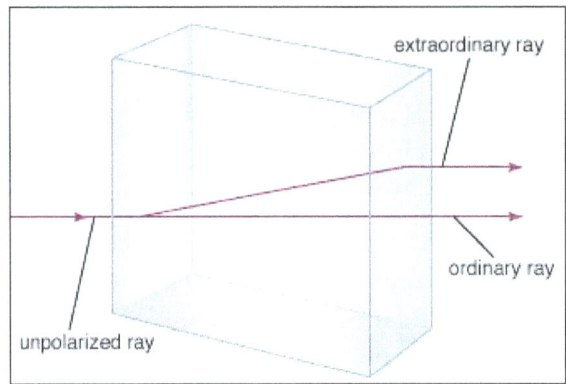

Double refraction showing two rays emerging when a single light ray
strikes a calcite crystal at a right angle to one face.

Maxwell's electromagnetic waves are transverse, with the electric and magnetic fields oscillating in directions perpendicular to the propagation direction. The fields are also perpendicular to one an-other, with the electric field direction, magnetic field direction, and propagation direction forming a right-handed coordinate system. For a wave with frequency f and wavelength λ (related by $\lambda f = c$) propagating in the positive x-direction, the fields are described mathematically by:

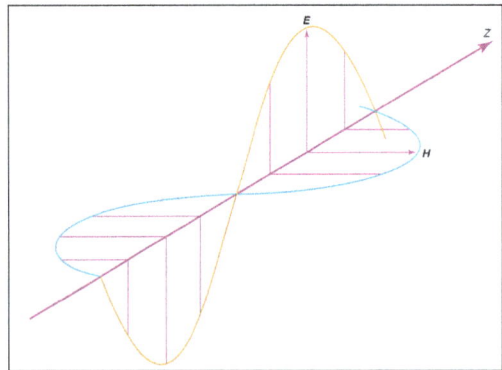

Electromagnetic wave, showing that electric field vector E and magnetic field vector H are in phase.

$$\vec{E}(x,t) = E_0 \cos\left(\frac{2\pi x}{\lambda} - 2\pi ft\right)\hat{y}$$

$$\vec{B}(x,t) = B_0 \cos\left(\frac{2\pi x}{\lambda} - 2\pi ft\right)\hat{z}.$$

The equations show that the electric and magnetic fields are in phase with each other; at any given point in space, they reach their maximum values, E_0 and B_0, at the same time. The amplitudes of the fields are not independent; Maxwell's equations show that $E_0 = cB_0$ for all electromagnetic waves in a vacuum.

In describing the orientation of the electric and magnetic fields of a light wave, it is common practice to specify only the direction of the electric field; the magnetic field direction then follows from the requirement that the fields are perpendicular to one another, as well as the direction of wave propagation. A linearly polarized wave has the property that the fields oscillate in fixed directions as the wave propagates. Other polarization states are possible. In a circularly polarized light wave, the electric and magnetic field vectors rotate about the propagation direction while maintaining fixed amplitudes. Elliptically polarized light refers to a situation intermediate between the linear and circular polarization states.

Unpolarized Light

The atoms on the surface of a heated filament, which generate light, act independently of one another. Each of their emissions can be approximately modeled as a short "wave train" lasting from about 10−9 to 10−8 second. The electromagnetic wave emanating from the filament is a superposition of these wave trains, each having its own polarization direction. The sum of the randomly oriented wave trains results in a wave whose direction of polarization changes rapidly and randomly. Such a wave is said to be unpolarized. All common sources of light, including the Sun, incandescent and fluorescent lights, and flames, produce unpolarized light. However, natural light is often partially polarized because of multiple scatterings and reflections.

Sources of Polarized Light

Polarized light can be produced in circumstances where a spatial orientation is defined. One example is synchrotron radiation, where highly energetic charged particles move in a magnetic field and emit polarized electromagnetic waves. There are many known astronomical sources of synchrotron radiation, including emission nebulae, supernova remnants, and active galactic nuclei; the polarization of astronomical light is studied in order to infer the properties of these sources.

Polarized lenses selectively block light of horizontal orientation—resulting in a dramatic decrease in glare, which consists mostly of light reflected off horizontal surfaces. Polarized lenses are commonly used in sunglasses, binoculars, telescopes, and cameras.

Natural light is polarized in passage through a number of materials, the most common being polaroid. Invented by the American physicist Edwin Land, a sheet of polaroid consists of long-chain hydrocarbon molecules aligned in one direction through a heat-treatment process. The molecules preferentially absorb any light with an electric field parallel to the alignment direction. The light emerging from a polaroid is linearly polarized with its electric field perpendicular to the alignment direction. Polaroid is used in many applications, including sunglasses and camera filters, to remove reflected and scattered light.

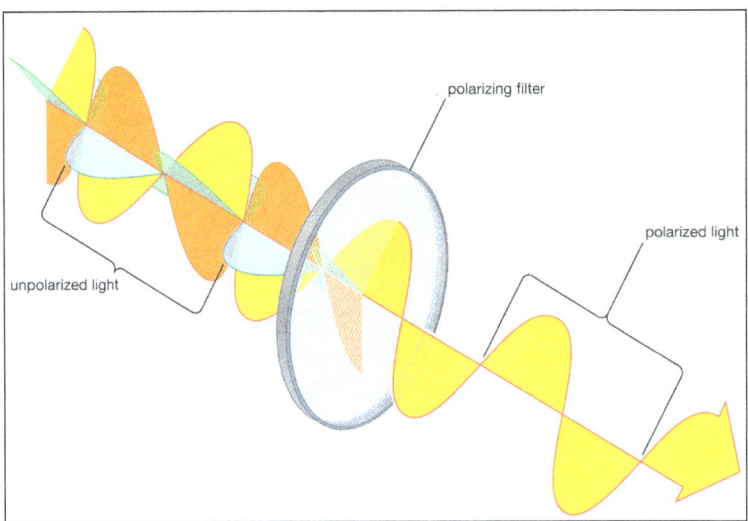

A polarizing filter has its molecules all aligned in the same direction. Light waves with the same orientation as the filter are absorbed by the molecules' vibrations, thereby reducing the intensity of the light passing through the filter.

In 1808 the French physicist Étienne-Louis Malus discovered that, when natural light reflects off a nonmetallic surface, it is partially polarized. The degree of polarization depends on the angle of incidence and the index of refraction of the reflecting material. At one extreme, when the tangent of the incident angle of light in air equals the index of refraction of the reflecting material, the reflected light is 100 percent linearly polarized; this is known as Brewster's law (after its discoverer, the Scottish physicist David Brewster). The direction of polarization is parallel to the reflecting surface. Because daytime glare typically originates from reflections off horizontal surfaces such as roads and water, polarizing filters are often used in sunglasses to remove horizontally polarized light, hence selectively removing glare.

The scattering of unpolarized light by very small objects, with sizes much less than the wavelength of the light (called Rayleigh scattering, after the English scientist Lord Rayleigh), also produces a partial polarization. When sunlight passes through Earth's atmosphere, it is scattered by air molecules. The scattered light that reaches the ground is partially linearly polarized, the extent of its polarization depending on the scattering angle. Because human eyes are not sensitive to the polarization of light, this effect generally goes unnoticed. However, the eyes of many insects are responsive to polarization properties, and they use the relative polarization of ambient sky light as a navigational tool. A common camera filter employed to reduce background light in bright sunshine is a simple linear polarizer designed to reject Rayleigh scattered light from the sky.

Polarization effects are observable in optically anisotropic materials (in which the index of refraction varies with polarization direction) such as birefringent crystals and some biological structures and in optically active materials. Technological applications include polarizing microscopes, liquid crystal displays, and optical instrumentation for materials testing.

Energy Transport

The transport of energy by light plays a critical role in life. About 1022 joules of solar radiant energy reaches Earth each day. Perhaps half of that energy reaches Earth's surface, the rest being absorbed or scattered in the atmosphere. In turn, Earth continuously reradiates electromagnetic energy (predominantly in the infrared). Together, these energy-transport processes determine Earth's energy balance, setting its average temperature and driving its global weather patterns. The transformation of solar energy into chemical energy by photosynthesis in plants maintains life on Earth. The fossil fuels that power industrial society—natural gas, petroleum, and coal—are ultimately stored organic forms of solar energy deposited on Earth millions of years ago.

The electromagnetic-wave model of light accounts naturally for the origin of energy transport. In an electromagnetic wave, energy is stored in the electric and magnetic fields; as the fields propagate at the speed of light, the energy content is transported. The proper measure of energy transport in an electromagnetic wave is its irradiance, or intensity, which equals the rate at which energy passes a unit area oriented perpendicular to the direction of propagation. The time-averaged irradiance I for a harmonic electromagnetic wave is related to the amplitudes of the electric and magnetic fields: $I = \varepsilon_0 c^2 E_0 B_0 / 2$ watts per square metre.

The irradiance of sunlight at the top of Earth's atmosphere is about 1,350 watts per square metre; this factor is referred to as the solar constant. Considerable efforts have gone into developing technologies to transform this solar energy into directly usable thermal or electric energy.

Radiation Pressure

In addition to carrying energy, light transports momentum and is capable of exerting mechanical forces on objects. When an electromagnetic wave is absorbed by an object, the wave exerts a pressure (P) on the object that equals the wave's irradiance (I) divided by the speed of light (c): $P = I/c$ newtons per square metre.

Most natural light sources exert negligibly small forces on objects; this subtle effect was first demonstrated in 1903 by the American physicists Ernest Fox Nichols and Gordon Hull. However, radiation pressure is consequential in a number of astronomical settings. Perhaps most important, the equilibrium conditions of stellar structure are determined largely by the opposing forces of gravitational attraction on the one hand and radiation pressure and thermal pressure on the other. The outward force of the light escaping the core of a star, working with thermal pressure, acts to balance the inward gravitational forces on the outer layers of the star. Another, visually dramatic, example of radiation pressure is the formation of cometary tails, in which dust particles released by cometary nuclei are pushed by solar radiation into characteristic trailing patterns.

Terrestrial applications of radiation pressure became feasible with the advent of lasers in the 1960s. In part because of the small diameters of their output beams and the excellent focusing properties of the beams, laser intensities are generally orders of magnitude larger than the intensities of natural light sources. On the largest scale, the most powerful laser systems are designed to compress and heat target materials in nuclear fusion inertial confinement schemes. The radiation forces from table-top laser systems are used to manipulate atoms and microscopic objects. The techniques of laser cooling and trapping, pioneered by the Nobelists Steven Chu, William Phillips, and Claude Cohen-Tannoudji, slow a gas of atoms in an "optical molasses" of intersecting laser beams. Temperatures below $10-6$ K (one-millionth of a degree above absolute zero) have been achieved. "Optical tweezers" is a related technique in which a tightly focused laser beam exerts a radiation force large enough to deflect, guide, and trap micron-sized objects ranging from dielectric spheres to biological samples such as viruses, single living cells, and organelles within cells.

Interactions of Light with Matter

The transmission of light through a piece of glass, the reflections and refractions of light in a raindrop, and the scattering of sunlight in Earth's atmosphere are examples of interactions of light with matter. On an atomic scale, these interactions are governed by the quantum mechanical natures of matter and light, but many are adequately explained by the interactions of classical electromagnetic radiation with charged particles.

A detailed presentation of the classical model of an interaction between an electromagnetic wave and an atom can be found in the article electromagnetic radiation. In brief, the electric and magnetic fields of the wave exert forces on the bound electrons of the atom, causing them to oscillate at the frequency of the wave. Oscillating charges are sources of electromagnetic radiation; the oscillating electrons radiate waves at the same frequency as the incoming fields. This constitutes the microscopic origin of the scattering of an electromagnetic wave. The electrons initially absorb energy from the incoming wave as they are set in motion, and they redirect that energy in the form of scattered light of the same frequency.

Through interference effects, the superposition of the reradiated waves from all of the participating atoms determines the net outcome of the scattering interactions. Two examples illustrate this point. As a light beam passes through transparent glass, the reradiated waves within the glass interfere destructively in all directions except the original propagation direction of the beam, resulting in little or no light's being scattered out of the original beam. Therefore, the light advances without loss through the glass. When sunlight passes through Earth's upper atmosphere, on the other hand, the reradiated waves generated by the gaseous molecules do not suffer destructive interference, so that a significant amount of light is scattered in many directions. The outcomes of these two scattering interactions are quite different, primarily because of differences in the densities of the scatterers. Generally, when the mean spacing between scatterers is significantly less than the wavelength of the light (as in glass), destructive interference effects significantly limit the amount of lateral scattering; when the mean spacing is greater than, or on the order of, the wavelength and the scatterers are randomly distributed in space (as in the upper atmosphere), interference effects do not play a significant role in the lateral scattering.

Lord Rayleigh's analysis in 1871 of the scattering of light by atoms and molecules in the atmosphere showed that the intensity of the scattered light increases as the fourth power of its frequency; this strong dependence on frequency explains the colour of the sunlit sky. Being at the high-frequency end of the visible spectrum, blue light is scattered far more by air molecules than the lower-frequency colours; the sky appears blue. On the other hand, when sunlight passes through a long column of air, such as at sunrise or sunset, the high-frequency components are selectively scattered out of the beam and the remaining light appears reddish.

Nonlinear Interactions

The interactions of light waves with matter become progressively richer as intensities are increased. The field of nonlinear optics describes interactions in which the response of the atomic oscillators is no longer simply proportional to the intensity of the incoming light wave. Nonlinear optics has many significant applications in communications and photonics, information processing, schemes for optical computing and storage, and spectroscopy.

Nonlinear effects generally become observable in a material when the strength of the electric field in the light wave is appreciable in comparison with the electric fields within the atoms of the material. Laser sources, particularly pulsed sources, easily achieve the required light intensities for this regime. Nonlinear effects are characterized by the generation of light with frequencies differing from the frequency of the incoming light beam. Classically, this is understood as resulting from the large driving forces of the electric fields of the incoming wave on the atomic oscillators. As an illustration, consider second harmonic generation, the first nonlinear effect observed in a crystal. When high-intensity light of frequency f passes through an appropriate nonlinear crystal (quartz was used in the first observations), a fraction of that light is converted to light of frequency 2f. Higher harmonics can also be generated with appropriate media, as well as combinations of frequencies when two or more light beams are used as input.

Quantum Theory of Light

By the end of the 19th century, the battle over the nature of light as a wave or a collection of particles seemed over. James Clerk Maxwell's synthesis of electric, magnetic, and optical phenomena and the discovery by Heinrich Hertz of electromagnetic waves were theoretical and experimental triumphs of the first order. Along with Newtonian mechanics and thermodynamics, Maxwell's electromagnetism took its place as a foundational element of physics. However, just when everything seemed to be settled, a period of revolutionary change was ushered in at the beginning of the 20th century. A new interpretation of the emission of light by heated objects and new experimental methods that opened the atomic world for study led to a radical departure from the classical theories of Newton and Maxwell—quantum mechanics was born. Once again the question of the nature of light was reopened.

Principal Historical Developments

Blackbody Radiation

Blackbody radiation refers to the spectrum of light emitted by any heated object; common examples include the heating element of a toaster and the filament of a light bulb. The spectral intensity of blackbody radiation peaks at a frequency that increases with the temperature of the emitting body: room temperature objects (about 300 K) emit radiation with a peak intensity in the far infrared; radiation from toaster filaments and light bulb filaments (about 700 K and 2,000 K, respectively) also peak in the infrared, though their spectra extend progressively into the visible; while the 6,000 K surface of the Sun emits blackbody radiation that peaks in the centre of the visible range. In the late 1890s, calculations of the spectrum of blackbody radiation based on classical electromagnetic theory and thermodynamics could not duplicate the results of careful measurements. In fact, the calculations predicted the absurd result that, at any temperature, the spectral intensity increases without limit as a function of frequency.

In 1900 the German physicist Max Planck succeeded in calculating a blackbody spectrum that matched experimental results by proposing that the elementary oscillators at the surface of any object (the detailed structure of the oscillators was not relevant) could emit and absorb electromagnetic radiation only in discrete packets, with the energy of a packet being directly proportional to the frequency of the radiation, E = hf. The constant of proportionality, h, which Planck determined by comparing his theoretical results with the existing experimental data, is now called Planck's constant and has the approximate value 6.626×10^{-34} joule·second.

Photons

Planck did not offer a physical basis for his proposal; it was largely a mathematical construct needed to match the calculated blackbody spectrum to the observed spectrum. In 1905 Albert Einstein gave a ground-breaking physical interpretation to Planck's mathematics when he proposed that electromagnetic radiation itself is granular, consisting of quanta, each with an energy hf. He based his conclusion on thermodynamic arguments applied to a radiation field that obeys Planck's radiation law. The term photon, which is now applied to the energy quantum of light, was later coined by the American chemist Gilbert N. Lewis.

Einstein supported his photon hypothesis with an analysis of the photoelectric effect, a process, discovered by Hertz in 1887, in which electrons are ejected from a metallic surface illuminated by light. Detailed measurements showed that the onset of the effect is determined solely by the frequency of the light and the makeup of the surface and is independent of the light intensity. This behaviour was puzzling in the context of classical electromagnetic waves, whose energies are proportional to intensity and independent of frequency. Einstein supposed that a minimum amount of energy is required to liberate an electron from a surface—only photons with energies greater than this minimum can induce electron emission. This requires a minimum light frequency, in agreement with experiment. Einstein's prediction of the dependence of the kinetic energy of the ejected electrons on the light frequency, based on his photon model, was experimentally verified by the American physicist Robert Millikan in 1916.

In 1922 American Nobelist Arthur Compton treated the scattering of X-rays from electrons as a set of collisions between photons and electrons. Adapting the relation between momentum and energy for a classical electromagnetic wave to an individual photon, $p = E/c = hf/c = h/\lambda$, Compton used the conservation laws of momentum and energy to derive an expression for the wavelength shift of scattered X-rays as a function of their scattering angle. His formula matched his experimental findings, and the Compton effect, as it became known, was considered further convincing evidence for the existence of particles of electromagnetic radiation.

The energy of a photon of visible light is very small, being on the order of 4×10^{-19} joule. A more convenient energy unit in this regime is the electron volt (eV). One electron volt equals the energy gained by an electron when its electric potential is changed by one volt: 1 eV = 1.6×10^{-19} joule. The spectrum of visible light includes photons with energies ranging from about 1.8 eV (red light) to about 3.1 eV (violet light). Human vision cannot detect individual photons, although, at the peak of its spectral response (about 510 nm, in the green), the dark-adapted eye comes close. Under normal daylight conditions, the discrete nature of the light entering the human eye is completely obscured by the very large number of photons involved. For example, a standard 100-watt light bulb emits on the order of 10^{20} photons per second; at a distance of 10 metres from the bulb, perhaps 10^{11} photons per second will enter a normally adjusted pupil of a diameter of 2 mm.

Photons of visible light are energetic enough to initiate some critically important chemical reactions, most notably photosynthesis through absorption by chlorophyll molecules. Photovoltaic systems are engineered to convert light energy to electric energy through the absorption of visible photons by semiconductor materials. More-energetic ultraviolet photons (4 to 10 eV) can initiate photochemical reactions such as molecular dissociation and atomic and molecular ionization. Modern methods for detecting light are based on the response of materials to individual photons. Photoemissive detectors, such as photomultiplier tubes, collect electrons emitted by the photoelectric effect; in photoconductive detectors the absorption of a photon causes a change in the conductivity of a semiconductor material.

A number of subtle influences of gravity on light, predicted by Einstein's general theory of relativity, are most easily understood in the context of a photon model of light and are presented here. (However, note that general relativity is not itself a theory of quantum physics.).

Through the famous relativity equation $E = mc^2$, a photon of frequency f and energy $E = hf$ can be considered to have an effective mass of $m = hf/c^2$. Note that this effective mass is distinct from the "rest mass" of a photon, which is zero. General relativity predicts that the path of light is deflected in the gravitational field of a massive object; this can be somewhat simplistically understood as resulting from a gravitational attraction proportional to the effective mass of the photons. In addition, when light travels toward a massive object, its energy increases, and its frequency thus increases (gravitational blueshift). Gravitational redshift describes the converse situation where light traveling away from a massive object loses energy and its frequency decreases.

Quantum Mechanics

The first two decades of the 20th century left the status of the nature of light confused. That light is a wave phenomenon was indisputable: there were countless examples of interference effects—the signature of waves—and a well-developed electromagnetic wave theory. However, there was also undeniable evidence that light consists of a collection of particles with well-defined energies and momenta. This paradoxical wave-particle duality was soon seen to be shared by all elements of the material world.

In 1923 the French physicist Louis de Broglie suggested that wave-particle duality is a feature common to light and all matter. In direct analogy to photons, de Broglie proposed that electrons with momentum p should exhibit wave properties with an associated wavelength $\lambda = h/p$. Four years later, de Broglie's hypothesis of matter waves, or de Broglie waves, was experimentally confirmed by Clinton Davisson and Lester Germer at Bell Laboratories with their observation of electron diffraction effects.

A radically new mathematical framework for describing the microscopic world, incorporating de Broglie's hypothesis, was formulated in 1926–27 by the German physicist Werner Heisenberg and the Austrian physicist Erwin Schrödinger, among others. In quantum mechanics, the dominant theory of 20th-century physics, the Newtonian notion of a classical particle with a well-defined trajectory is replaced by the wave function, a nonlocalized function of space and time. The interpretation of the wave function, originally suggested by the German physicist Max Born, is statistical—the wave function provides the means for calculating the probability of finding a particle at any point in space. When a measurement is made to detect a particle, it always appears as pointlike, and its position immediately after the measurement is well defined. But before a measurement is made, or between successive measurements, the particle's position is not well defined; instead, the state of the particle is specified by its evolving wave function.

The quantum mechanics embodied in the 1926–27 formulation is nonrelativistic—that is, it applies only to particles whose speeds are significantly less than the speed of light. The quantum mechanical description of light was not fully realized until the late 1940s. However, light and matter share a common central feature—a complementary relation between wave and particle aspects—that can be illustrated without resorting to the formalisms of relativistic quantum mechanics.

Wave-particle Duality

The same interference pattern demonstrated in Young's double-slit experiment is produced when a beam of matter, such as electrons, impinges on a double-slit apparatus. Concentrating on light, the interference pattern clearly demonstrates its wave properties. But what of its particle properties? Can an individual photon be followed through the two-slit apparatus, and if so, what is the origin of the resulting interference pattern? The superposition of two waves, one passing through each slit, produces the pattern in Young's apparatus. Yet, if light is considered a collection of particle-like photons, each can pass only through one slit or the other. Soon after Einstein's photon hypothesis in 1905, it was suggested that the two-slit interference pattern might be caused by the interaction of photons that passed through different slits. This interpretation was ruled out in 1909 when the English physicist Geoffrey Taylor reported a diffraction pattern in the shadow of a needle recorded on a photographic plate exposed to a very weak light source, weak enough that only one photon could be present in the apparatus at any one time. Photons were not interfering with one another; each photon was contributing to the diffraction pattern on its own.

In modern versions of this two-slit interference experiment, the photographic plate is replaced with a detector that is capable of recording the arrival of individual photons. Each photon arrives whole and intact at one point on the detector. It is impossible to predict the arrival position of any one photon, but the cumulative effect of many independent photon impacts on the detector results in the gradual buildup of an interference pattern. The magnitude of the classical interference pattern at any one point is therefore a measure of the probability of any one photon's arriving at that point. The interpretation of this seemingly paradoxical behaviour (shared by light and matter), which is in fact predicted by the laws of quantum mechanics, has been debated by the scientific community since its discovery more than 100 years ago. The American physicist Richard Feynman summarized the situation in 1965:

> We choose to examine a phenomenon which is impossible, absolutely impossible, to explain in any classical way, and which has in it the heart of quantum mechanics. In reality, it contains the only mystery.

In a wholly unexpected fashion, quantum mechanics resolved the long wave-particle debate over the nature of light by rejecting both models. The behaviour of light cannot be fully accounted for by a classical wave model or by a classical particle model. These pictures are useful in their respective regimes, but ultimately they are approximate, complementary descriptions of an underlying reality that is described quantum mechanically.

Quantum Optics

Quantum optics, the study and application of the quantum interactions of light with matter, is an active and expanding field of experiment and theory. Progress in the development of light sources and detection techniques since the early 1980s has allowed increasingly sophisticated optical tests of the foundations of quantum mechanics. Basic quantum effects such as single photon interference, along with more esoteric issues such as the meaning of the measurement process, have been more clearly elucidated. Entangled states of two or more photons with highly correlated properties (such as polarization direction) have been generated and used to test the fundamental issue of nonlocality in quantum mechanics. Novel technological applications of quantum optics are also under study, including quantum cryptography and quantum computing.

Emission and Absorption Processes

Bohr Model

That materials, when heated in flames or put in electrical discharges, emit light at well-defined and characteristic frequencies was known by the mid-19th century. The study of the emission and absorption spectra of atoms was crucial to the development of a successful theory of atomic structure. Attempts to describe the origin of the emission and absorption lines (i.e., the frequencies of emission and absorption) of even the simplest atom, hydrogen, in the framework of classical mechanics and electromagnetism failed miserably. Then, in 1913, Danish physicist Niels Bohr proposed a model for the hydrogen atom that succeeded in explaining the regularities of its spectrum. In what is known as the Bohr atomic model, the orbiting electrons in an atom are found in only certain allowed "stationary states" with well-defined energies. An atom can absorb or emit one photon when an electron makes a transition from one stationary state, or energy level, to another. Conservation of energy determines the energy of the photon and thus the frequency of the emitted or absorbed light. Though Bohr's model was superseded by quantum mechanics, it still offers a useful, though simplistic, picture of atomic transitions.

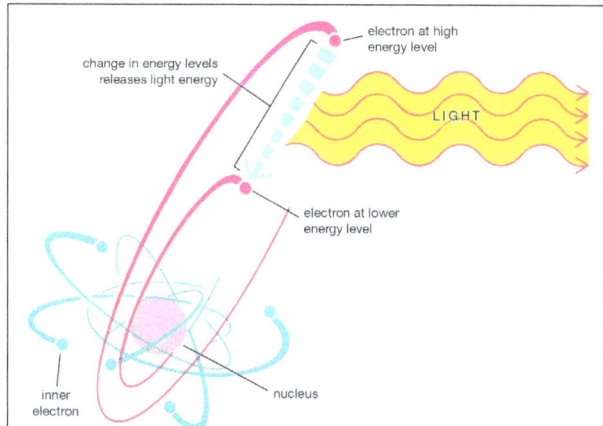

emission of lightModern theory explains the emission of light by matter in terms of electronic energy levels. An electron of relatively high energy may jump to a condition of lower energy, giving off the energy difference as electromagnetic radiation.

Spontaneous Emission

When an isolated atom is excited into a high-energy state, it generally remains in the excited state for a short time before emitting a photon and making a transition to a lower energy state. This fundamental process is called spontaneous emission. The emission of a photon is a probabilistic event; that is, the likelihood of its occurrence is described by a probability per unit time. For many excited states of atoms, the average time before the spontaneous emission of a photon is on the order of 10^{-9} to 10^{-8} second.

Stimulated Emission

The absorption of a photon by an atom is also a probabilistic event, with the probability per unit time being proportional to the intensity of the light falling on the atom. In 1917 Einstein, though not knowing the exact mechanisms for the emission and absorption of photons, showed through thermodynamic arguments that there must be a third type of radiative transition in an atom— stimulated emission. In stimulated emission the presence of photons with an appropriate energy triggers an atom in an excited state to emit a photon of identical energy and to make a transition to a lower state. As with absorption, the probability of stimulated emission is proportional to the intensity of the light bathing the atom. Einstein mathematically expressed the statistical nature of the three possible radiative transition routes (spontaneous emission, stimulated emission, and absorption) with the so-called Einstein coefficients and quantified the relations between the three processes. One of the early successes of quantum mechanics was the correct prediction of the numerical values of the Einstein coefficients for the hydrogen atom.

Einstein's description of the stimulated emission process showed that the emitted photon is identical in every respect to the stimulating photons, having the same energy and polarization, traveling in the same direction, and being in phase with those photons. Some 40 years after Einstein's work, the laser was invented, a device that is directly based on the stimulated emission process. (The acronym laser stands for "light amplification by stimulated emission of radiation.") Laser light, because of the underlying properties of stimulated emission, is highly monochromatic, directional, and coherent. Many modern spectroscopic techniques for probing atomic and molecular structure and dynamics, as well as innumerable technological applications, take advantage of these properties of laser light.

Quantum Electrodynamics

The foundations of a quantum mechanical theory of light and its interactions with matter were developed in the late 1920s and '30s by Paul Dirac, Werner Heisenberg, Pascual Jordan, Wolfgang Pauli, and others. The fully developed theory, called quantum electrodynamics (QED), is credited to the independent work of Richard Feynman, Julian S. Schwinger, and Tomonaga Shin'ichirō. QED describes the interactions of electromagnetic radiation with charged particles and the interactions of charged particles with one another. The electric and magnetic fields described in Maxwell's equations are quantized, and photons appear as excitations of those quantized fields. In QED, photons serve as carriers of electric and magnetic forces. For example, two identical charged particles electrically repel one another because they are exchanging what are called virtual photons. (Virtual photons cannot be directly detected; their existence violates the conservation laws of energy and momentum.) Photons can also be freely emitted by charged particles, in which case they are detectable as light. Though the mathematical complexities of QED are formidable, it is a highly successful theory that has now withstood decades of precise experimental tests. It is considered the prototype field theory in physics; great efforts have gone into adapting its core concepts and calculational approaches to the description of other fundamental forces in nature.

QED provides a theoretical framework for processes involving the transformations of matter into photons and photons into matter. In pair creation, a photon interacting with an atomic nucleus (to conserve momentum) disappears, and its energy is converted into an electron and a positron (a particle-antiparticle pair). In pair annihilation, an electron-positron pair disappears, and two high-energy photons are created. These processes are of central importance in cosmology—once again demonstrating that light is a primary component of the physical universe.

REFLECTION OF LIGHT

Reflection of Light is the process of sending back the light rays which falls on the surface of an object. The image formed due to reflection of an object on a plane mirror is at different places.

Light travels in a straight line. It can either be reflected or refracted.

Reflection of Light

The process through which light rays falling on the surface on an object are sent back is called reflection of light. Thus, when light falls on the surface of an object it sends back the light.

The objects having shiny or polished surface reflects more light compared to the objects having dull or unpolished surface. Silver metal is the best reflector of light. This is why plane mirror is made by depositing a thin layer of silver metal on one side of a plane glass sheet. The silver coating is protected by a red paint.

The straight line along which the light travels is called ray of light.

Regular Reflection and Diffuse Reflection of Light

In regular reflection, a parallel beam of incident light is reflected as a parallel beam in one direction. In this case , parallel incident rays remain parallel even after reflection and go only in one direction and it occurs from smooth surfaces like that of a plane mirror or highly polished metal surfaces. Thus, a plane mirror produces regular reflection of light. Since the angle of incidence and the angle of reflection are the same or equal, a beam of parallel rays falling on a smooth surface is reflected as a beam of parallel light rays in one direction only. It is explained below in the figure.

In diffuse reflection, a parallel beam of incident light is reflected in different directions. In this case, the parallel incident rays do not remain parallel after reflection, they are scattered in different directions. It is also known as irregular reflection or scattering and so, takes place from rough surfaces like that of paper, cardboard, chalk, table, chair, walls and unpolished metal objects. Since, the angle of incidence and angle of reflection are different, the parallel rays of light falling on a rough surface go in different directions as explained below in the figure.

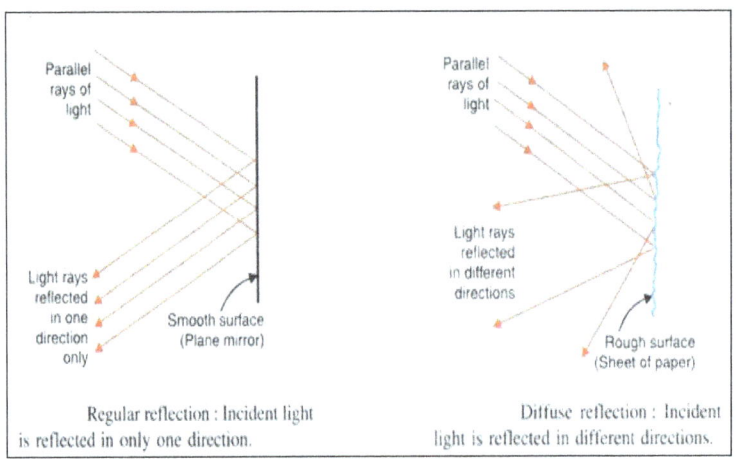

Regular reflection : Incident light is reflected in only one direction.

Diffuse reflection : Incident light is reflected in different directions.

Reflection of Light from Plane Mirror

Before understanding the laws of reflection of light, lets understand the meaning of some important terms such as, incident ray, reflected ray, point of incidence, normal (at the point of incidence), angle of incidence and angle of reflection.

- Incident ray: The ray of light falling on the surface of a mirror is called incident ray.

- Point of incidence: The point at which the incident ray falls on the mirror surface is called point of incidence.

- Reflected ray: The ray of light which is sent back by the mirror from the point of incidence is called reflected ray.

- Normal: A line perpendicular or at the right angle to the mirror surface at the point of incidence is called normal.

- Angle of incidence: The angle made by the incident ray with the normal is called angle of incidence.

- Angle of reflection: The angle made by the reflected ray with the normal at point of incidence is called angle of reflection.

Laws of Reflection of Light

The laws of reflection of light apply to both plane mirror as well as spherical mirror.

- First law of reflection: According to the first law, the incident ray, reflected ray and normal, all lie in the same plane.

- Second law of reflection: According to the second law, the angle of reflection is always equal to the angle of incidence.

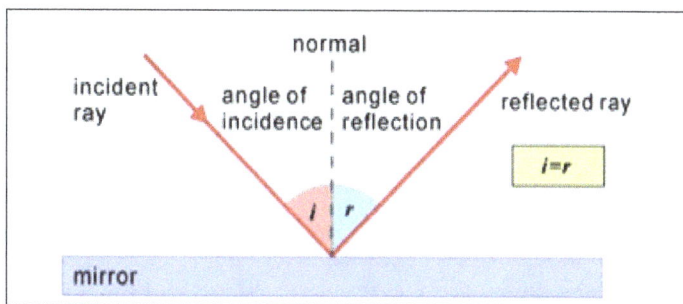

Also, it is important to note that when a ray of light falls normally on the surface of the mirror then the angle of incidence and the angle of reflection for such a ray of light will be zero. This ray of light will be reflected back along the same path.

Objects and Images

Anything which gives out light with off its own of reflected by it is called an object. For example, a bulb, a candle, a tree etc.

When the light rays coming from an object are reflected from a mirror then an optical appearance which is produced is called an image. For example, when we look into the mirror, we see the image of our face. Images are of two types, real image and virtual image.

Real image: The image which can be seen on screen is called real image.

Virtual image: The image which cannot be obtained on a screen is called virtual image.

Lateral inversion: When we stand in front of a mirror and lift our right hand than the image formed will lift its left hand. Therefore the right side of our body becomes the left side in its image and the left side of our body becomes the right side in its image in mirror.

The change of sides of an object in its mirror image is called lateral inversion. It happens due to reflection of light.

Formation of Image in a Plane Mirror

The nature of image formed by a plane mirror is:

- Virtual and erect.

- Size of image formed is equal to the size of object.

- Image is formed behind the mirror.

- Image is at same distance behind the mirror as the object is in front of the mirror.

- Image formed in plane mirror is laterally inverted.

Uses of plane mirror:

- Mirrors on our dressing table and bathrooms are plane mirrors and are used to see ourselves.

- They are fixed on the inside walls of jewellery shops to make them look big.

- They are fitted at blind turns on the roads so that the driver can see the vehicles coming from other side.

- Used in making periscopes.

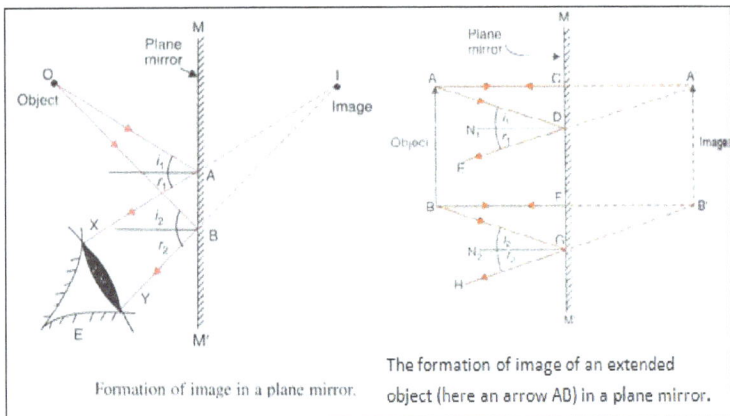

Formation of image in a plane mirror.

The formation of image of an extended object (here an arrow AB) in a plane mirror.

REFRACTION OF LIGHT

Refraction of light can be seen in many places in our everyday life. It makes objects under a water surface appear closer than they really are. It is what optical lenses are based on, allowing for instruments such as glasses, cameras, binoculars, microscopes, and the human eye. Refraction is also responsible for some natural optical phenomena including rainbows and mirages.

Law of Refraction

For light, the refractive index n of a material is more often used than the wave phase speed v in the material. They are, however, directly related through the speed of light in vacuum c as:

$$n = \frac{c}{v}.$$

In optics, therefore, the law of refraction is typically written as:

$$n_1 \sin \theta_1 = n_2 \sin \theta_2.$$

Refraction in a Water Surface

Refraction occurs when light goes through a water surface since water has a refractive index of

1.33 and air has a refractive index of about 1. Looking at a straight object, such as a pencil in the figure here, which is placed at a slant, partially in the water, the object appears to bend at the water's surface. This is due to the bending of light rays as they move from the water to the air. Once the rays reach the eye, the eye traces them back as straight lines (lines of sight). The lines of sight (shown as dashed lines) intersect at a higher position than where the actual rays originated. This causes the pencil to appear higher and the water to appear shallower than it really is.

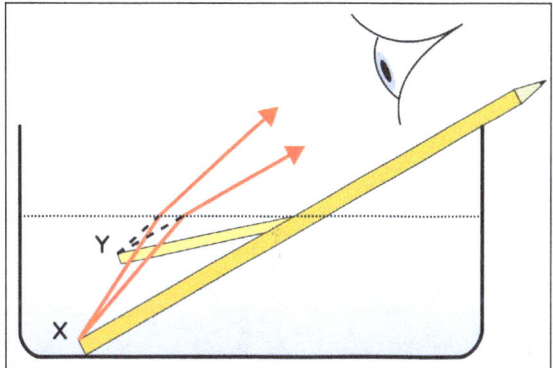

A pencil part immersed in water looks bent due to refraction: the light waves from X change direction and so seem to originate at Y.

The depth that the water appears to be when viewed from above is known as the *apparent depth*. This is an important consideration for spearfishing from the surface because it will make the target fish appear to be in a different place, and the fisher must aim lower to catch the fish. Conversely, an object above the water has a higher *apparent height* when viewed from below the water. The opposite correction must be made by an archer fish.

For small angles of incidence (measured from the normal, when sin θ is approximately the same as tan θ), the ratio of apparent to real depth is the ratio of the refractive indexes of air to that of water. But, as the angle of incidence approaches 90°, the apparent depth approaches zero, albeit reflection increases, which limits observation at high angles of incidence. Conversely, the apparent height approaches infinity as the angle of incidence (from below) increases, but even earlier, as the angle of total internal reflection is approached, albeit the image also fades from view as this limit is approached.

An image of the Golden Gate Bridge is refracted and bent by many differing three-dimensional drops of water.

Dispersion

Refraction is also responsible for rainbows and for the splitting of white light into a rainbow-spectrum as it passes through a glass prism. Glass has a higher refractive index than air. When a beam of white light passes from air into a material having an index of refraction that varies with frequency, a phenomenon known as dispersion occurs, in which different coloured components of the white light are refracted at different angles, i.e., they bend by different amounts at the interface, so that they become separated. The different colors correspond to different frequencies.

Atmospheric Refraction

The sun appears slightly flattened when close to the horizon due to refraction in the atmosphere.

The refractive index of air depends on the air density and thus vary with air temperature and pressure. Since the pressure is lower at higher altitudes, the refractive index is also lower, causing light rays to refract towards the earth surface when traveling long distances through the atmosphere. This shifts the apparent positions of stars slightly when they are close to the horizon and makes the sun visible before it geometrically rises above the horizon during a sunrise.

Heat haze in the engine exhaust above a diesel locomotive.

Mirage over a hot road.

Temperature variations in the air can also cause refraction of light. This can be seen as a heat haze when hot and cold air is mixed e.g. over a fire, in engine exhaust, or when opening a window on a cold day. This makes objects viewed through the mixed air appear to shimmer or move around randomly as the hot and cold air moves. This effect is also visible from normal variations in air temperature during a sunny day when using high magnification telephoto lenses and is often limiting the image quality in these cases. In a similar way, atmospheric turbulence gives rapidly varying distortions in the images of astronomical telescopes limiting the resolution of terrestrial telescopes not using adaptive optics or other techniques for overcoming these atmospheric distortions.

Air temperature variations close to the surface can give rise to other optical phenomena, such as mirages and Fata Morgana. Most commonly, air heated by a hot road on a sunny day deflects light approaching at a shallow angle towards a viewer. This makes the road appear reflecting, giving an illusion of water covering the road.

DIFFRACTION OF LIGHT

Single-slit Diffraction

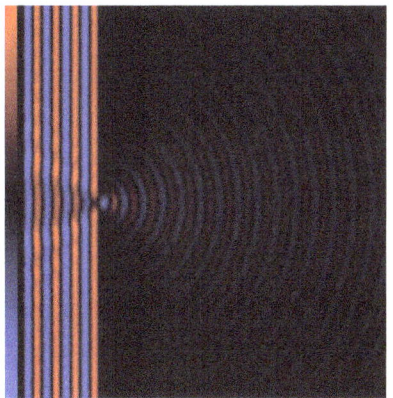

Diffraction of a scalar wave passing through a 1-wavelength-wide slit.

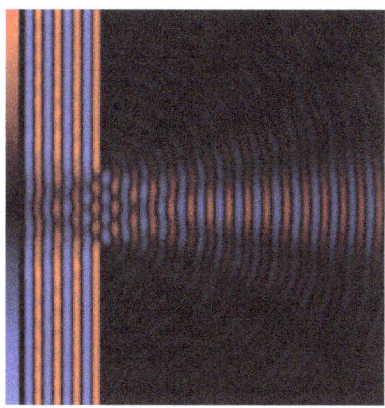

Diffraction of a scalar wave passing through a 4-wavelength-wide slit.

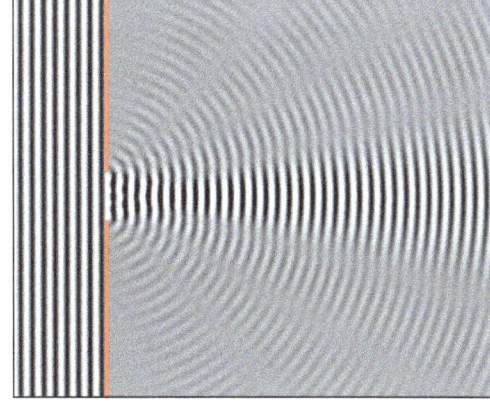

Numerical approximation of diffraction pattern from a slit of width four wavelengths with an incident plane wave. The main central beam, nulls, and phase reversals are apparent.

A long slit of infinitesimal width which is illuminated by light diffracts the light into a series of circular waves and the wavefront which emerges from the slit is a cylindrical wave of uniform intensity.

A slit which is wider than a wavelength produces interference effects in the space downstream of the slit. These can be explained by assuming that the slit behaves as though it has a large number of point sources spaced evenly across the width of the slit. The analysis of this system is simplified if we consider light of a single wavelength. If the incident light is coherent, these sources all have the same phase. Light incident at a given point in the space downstream of the slit is made up of contributions from each of these point sources and if the relative phases of these contributions vary by 2π or more, we may expect to find minima and maxima in the diffracted light. Such phase differences are caused by differences in the path lengths over which contributing rays reach the point from the slit.

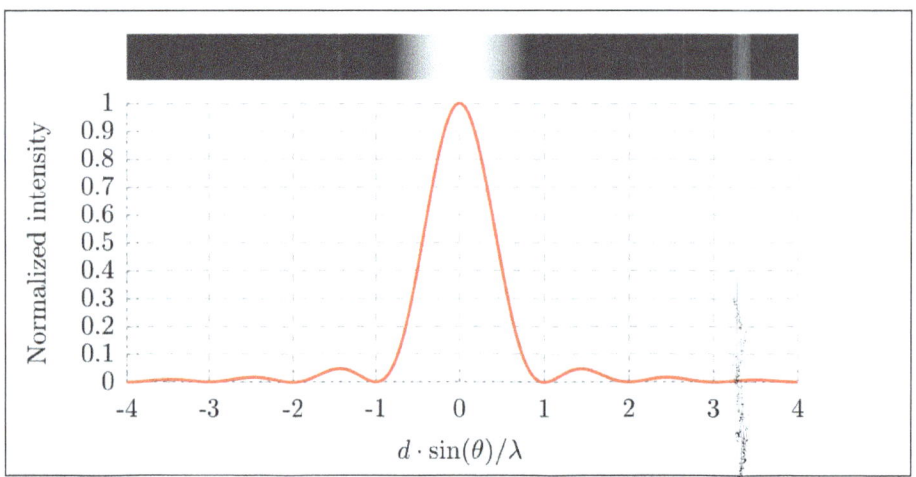

Graph and image of single-slit diffraction.

We can find the angle at which a first minimum is obtained in the diffracted light by the following reasoning. The light from a source located at the top edge of the slit interferes destructively with a source located at the middle of the slit, when the path difference between them is equal to $\lambda/2$. Similarly, the source just below the top of the slit will interfere destructively with the source located just below the middle of the slit at the same angle. We can continue this reasoning along the entire height of the slit to conclude that the condition for destructive interference for the entire slit is the same as the condition for destructive interference between two narrow slits a distance apart that is half the width of the slit. The path difference is approximately $\dfrac{d\sin(\theta)}{2}$ so that the minimum intensity occurs at an angle $\theta_{\min}$ given by:

$$d\sin\theta_{\min} = \lambda$$

where,

- d is the width of the slit,

- $\theta_{\min}$ is the angle of incidence at which the minimum intensity occurs, and

- λ is the wavelength of the light.

A similar argument can be used to show that if we imagine the slit to be divided into four, six, eight parts, etc., minima are obtained at angles θ_n given by:

$$d\sin\theta_n = n\lambda$$

where,

- n is an integer other than zero.

There is no such simple argument to enable us to find the maxima of the diffraction pattern. The intensity profile can be calculated using the Fraunhofer diffraction equation as:

$$I(\theta) = I_0 \operatorname{sinc}^2\left(\frac{d\pi}{\lambda}\sin\theta\right)$$

where,

- $I(\theta)$ is the intensity at a given angle,

- I_0 is the original intensity, and

- the unnormalized sinc function above is given by $\operatorname{sinc}(x) = \dfrac{\sin x}{x}$ if $x \neq 0$, and $\operatorname{sinc}(0) = 1$

This analysis applies only to the far field, that is, at a distance much larger than the width of the slit.

From the intensity profile above, if $d \ll \lambda$, the intensity will have little dependency on θ, hence the wavefront emerging from the slit would resemble a cylindrical wave of uniform intensity; If $d \ll \lambda$, only $\theta \approx 0$ would have appreciable intensity, hence the wavefront emerging from the slit would resemble that of geometrical optics.

Diffraction Grating

A diffraction grating is an optical component with a regular pattern. The form of the light diffracted by a grating depends on the structure of the elements and the number of elements present, but all gratings have intensity maxima at angles θ_m which are given by the grating equation:

$$d\left(\sin\theta_m + \sin\theta_i\right) = m\lambda$$

where,

- θ_i is the angle at which the light is incident,

- d is the separation of grating elements, and

- m is an integer which can be positive or negative.

The light diffracted by a grating is found by summing the light diffracted from each of the elements, and is essentially a convolution of diffraction and interference patterns.

The figure shows the light diffracted by 2-element and 5-element gratings where the grating spacings are the same; it can be seen that the maxima are in the same position, but the detailed structures of the intensities are different.

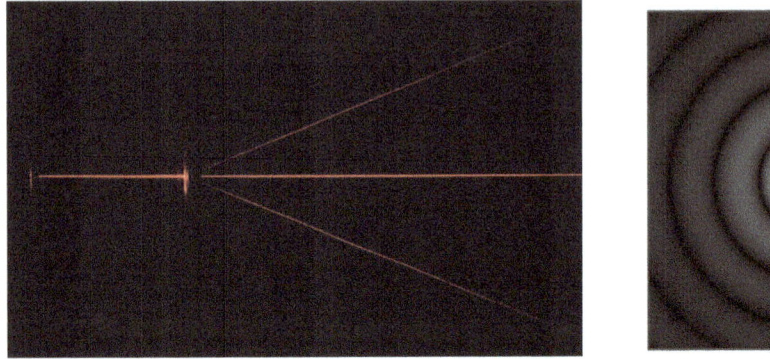

Diffraction of a red laser using a diffraction grating.

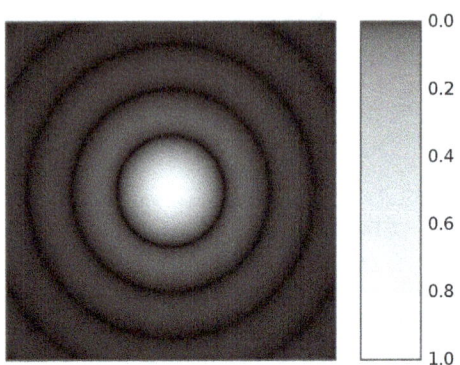

A computer-generated image of an Airy disk.

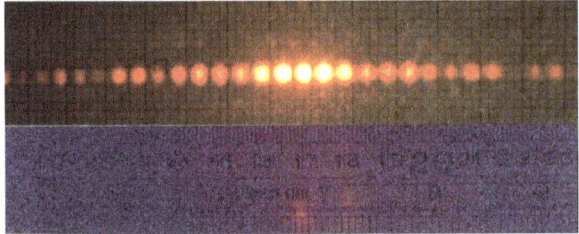

A diffraction pattern of a 633 nm laser through
a grid of 150 slits.

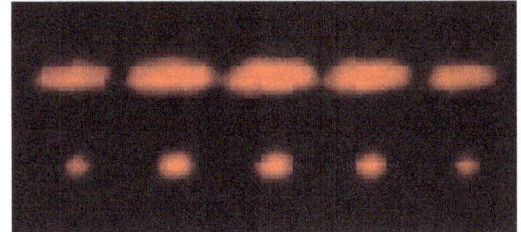

2-slit (top) and 5-slit diffraction
of red laser light.

Circular Aperture

The far-field diffraction of a plane wave incident on a circular aperture is often referred to as the Airy Disk. The variation in intensity with angle is given by:

$$I(\theta) = I_0 \left(\frac{2J_1(ka\sin\theta)}{ka\sin\theta} \right)^2,$$

where a is the radius of the circular aperture, k is equal to $2\pi/\lambda$ and J_1 is a Bessel function. The smaller the aperture, the larger the spot size at a given distance, and the greater the divergence of the diffracted beams.

General Aperture

The wave that emerges from a point source has amplitude ψ at location r that is given by the solution of the frequency domain wave equation for a point source (The Helmholtz Equation),

$$\nabla^2\psi + k^2\psi = \delta(\mathbf{r})$$

where $\delta(\mathbf{r})$ is the 3-dimensional delta function. The delta function has only radial dependence, so the Laplace operator (a.k.a. scalar Laplacian) in the spherical coordinate system simplifies to:

$$\nabla^2\psi = \frac{1}{r}\frac{\partial^2}{\partial r^2}(r\psi)$$

By direct substitution, the solution to this equation can be readily shown to be the scalar Green's function, which in the spherical coordinate system (and using the physics time convention $e^{-i\omega t}$) is:

$$\psi(r) = \frac{e^{ikr}}{4\pi r}$$

This solution assumes that the delta function source is located at the origin. If the source is located at an arbitrary source point, denoted by the vector $\mathbf{r}'$ and the field point is located at the point $\mathbf{r}$, then we may represent the scalar Green's function (for arbitrary source location) as:

$$\psi(\mathbf{r}\,|\,\mathbf{r}') = \frac{e^{ik|\mathbf{r}-\mathbf{r}'|}}{4\pi\,|\mathbf{r}-\mathbf{r}'|}$$

Therefore, if an electric field, $E_{inc}(x,y)$ is incident on the aperture, the field produced by this aperture distribution is given by the surface integral:

$$\Psi(r) \propto \iint\limits_{aperture} E_{inc}(x',y') \frac{e^{ik|\mathbf{r}-\mathbf{r}'|}}{4\pi\,|\mathbf{r}-\mathbf{r}'|} dx'dy',$$

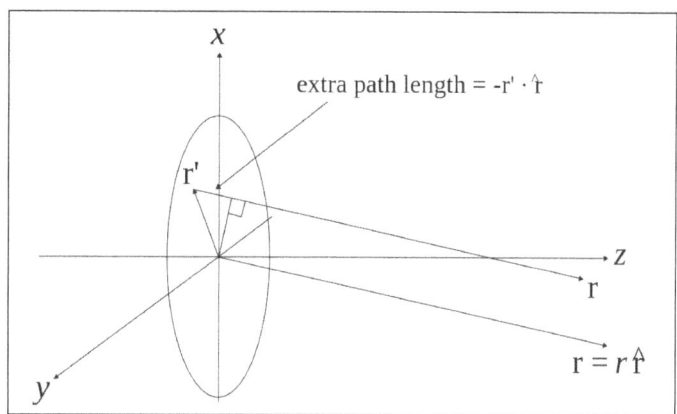

On the calculation of Fraunhofer region fields.

where the source point in the aperture is given by the vector:

$$\mathbf{r}' = x'\hat{\mathbf{x}} + y'\hat{\mathbf{y}}$$

In the far field, wherein the parallel rays approximation can be employed, the Green's function,

$$\psi(\mathbf{r}\,|\,\mathbf{r}') = \frac{e^{ik|\mathbf{r}-\mathbf{r}'|}}{4\pi\,|\mathbf{r}-\mathbf{r}'|}$$

simplifies to:

$$\psi(\mathbf{r}\,|\,\mathbf{r}') = \frac{e^{ikr}}{4\pi r} e^{-ik(\mathbf{r}'\cdot\hat{\mathbf{r}})}$$

as can be seen in the figure to the right (click to enlarge).

The expression for the far-zone (Fraunhofer region) field becomes:

$$\Psi(r) \propto \frac{e^{ikr}}{4\pi r} \iint\limits_{\text{aperture}} E_{\text{inc}}(x',y')e^{-ik(\mathbf{r}'\cdot\hat{\mathbf{r}})}\,dx'dy',$$

Now, since:

$$\mathbf{r}' = x'\hat{\mathbf{x}} + y'\hat{\mathbf{y}}$$

and

$$\hat{\mathbf{r}} = \sin\theta\cos\phi\hat{\mathbf{x}} + \sin\theta\,\sin\phi\hat{\mathbf{y}} + \cos\theta\hat{\mathbf{z}}$$

the expression for the Fraunhofer region field from a planar aperture now becomes,

$$\Psi(r) \propto \frac{e^{ikr}}{4\pi r} \iint\limits_{\text{aperture}} E_{\text{inc}}(x',y')e^{-ik\sin\theta(\cos\phi x'+\sin\phi y')}\,dx'dy'$$

Letting,

$$k_x = k\sin\theta\cos\phi$$

and

$$k_y = k\sin\theta\sin\phi$$

the Fraunhofer region field of the planar aperture assumes the form of a Fourier transform:

$$\Psi(r) \propto \frac{e^{ikr}}{4\pi r} \iint\limits_{\text{aperture}} E_{\text{inc}}(x',y')e^{-i(k_x x'+k_y y')}\,dx'dy',$$

In the far-field / Fraunhofer region, this becomes the spatial Fourier transform of the aperture distribution. Huygens' principle when applied to an aperture simply says that the far-field diffraction pattern is the spatial Fourier transform of the aperture shape, and this is a direct by-product of using the parallel-rays approximation, which is identical to doing a plane wave decomposition of the aperture plane fields.

Propagation of a Laser Beam

The way in which the beam profile of a laser beam changes as it propagates is determined by diffraction. When the entire emitted beam has a planar, spatially coherent wave front, it approximates Gaussian beam profile and has the lowest divergence for a given diameter. The smaller the output beam, the quicker it diverges. It is possible to reduce the divergence of a laser beam by first expanding it with one convex lens, and then collimating it with a second convex lens whose focal point is coincident with that of the first lens. The resulting beam has a larger diameter, and hence a lower divergence. Divergence of a laser beam may be reduced below the diffraction of a Gaussian beam or even reversed to convergence if the refractive index of the propagation media increases with the light intensity. This may result in a self-focusing effect.

When the wave front of the emitted beam has perturbations, only the transverse coherence length (where the wave front perturbation is less than 1/4 of the wavelength) should be considered as a Gaussian beam diameter when determining the divergence of the laser beam. If the transverse coherence length in the vertical direction is higher than in horizontal, the laser beam divergence will be lower in the vertical direction than in the horizontal.

Diffraction-limited Imaging

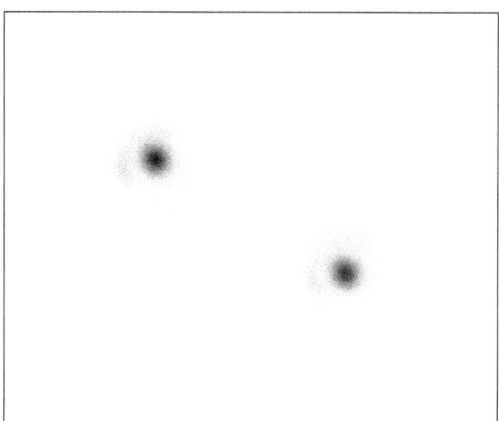

The Airy disk around each of the stars from the 2.56 m telescope aperture can be seen in this *lucky image* of the binary star zeta Boötis.

The ability of an imaging system to resolve detail is ultimately limited by diffraction. This is because a plane wave incident on a circular lens or mirror is diffracted as described above. The light is not focused to a point but forms an Airy disk having a central spot in the focal plane with radius to first null of:

$$d = 1.22 \lambda N$$

where λ is the wavelength of the light and N is the f-number (focal length divided by diameter) of the imaging optics. In object space, the corresponding angular resolution is:

$$\sin \theta = 1.22 \frac{\lambda}{D},$$

where D is the diameter of the entrance pupil of the imaging lens (e.g., of a telescope's main mirror).

Two point sources will each produce an Airy pattern – see the photo of a binary star. As the point sources move closer together, the patterns will start to overlap, and ultimately they will merge to form a single pattern, in which case the two point sources cannot be resolved in the image. The Rayleigh criterion specifies that two point sources can be considered to be resolvable if the separation of the two images is at least the radius of the Airy disk, i.e. if the first minimum of one coincides with the maximum of the other.

Thus, the larger the aperture of the lens, and the smaller the wavelength, the finer the resolution of an imaging system. This is why telescopes have very large lenses or mirrors, and why optical microscopes are limited in the detail which they can see.

Speckle Patterns

The speckle pattern which is seen when using a laser pointer is another diffraction phenomenon. It is a result of the superposition of many waves with different phases, which are produced when a laser beam illuminates a rough surface. They add together to give a resultant wave whose amplitude, and therefore intensity, varies randomly.

Babinet's Principle

Babinet's Principle is a useful theorem stating that the diffraction pattern from an opaque body is identical to that from a hole of the same size and shape, but with differing intensities. This means that the interference conditions of a single obstruction would be the same as that of a single slit.

References

- The Feynman Lectures on Physics Vol. I Ch. 30: Diffraction". www.feynmanlectures.caltech.edu. Retrieved 2019-04-25

- Unpolarized-light, light, science: britannica.com, Retrieved 29 January, 2019

- Andrew Norton (2000). Dynamic fields and waves of physics. CRC Press. p. 102. ISBN 978-0-7503-0719-2

- Reflection-of-light-1455278459-1, general-knowledge: jagranjosh.com, Retrieved 1 February, 2019

- "Refraction". RP Photonics Encyclopedia. RP Photonics Consulting gmbh, Dr. Rüdiger Paschotta. Retrieved

4

Sound

Sound is a longitudinal wave that requires material medium such as solid, liquid or gas, to propagate. The reflection of direct sound that can be heard after a short interval of time is referred to as echo. This chapter has been carefully written to provide in-depth knowledge of the varied facets of sound.

Sound is produced when something vibrates. The vibrating body causes the medium (water, air, etc.) around it to vibrate. Vibrations in air are called traveling longitudinal waves, which we can hear. Sound waves consist of areas of high and low pressure called compressions and rarefactions, respectively. Shown in the diagram below is a traveling wave. The shaded bar above it represents the varying pressure of the wave. Lighter areas are low pressure (rarefactions) and darker areas are high pressure (compressions). One wavelength of the wave is highlighted in red. This pattern repeats indefinitely. The wavelength of voice is about one meter long. The wavelength and the speed of the wave determine the pitch, or frequency of the sound. Wavelength, frequency, and speed are related by the equation speed = frequency * wavelength. Since sound travels at 343 meters per second at standard temperature and pressure (STP), speed is a constant. Thus, frequency is determined by speed / wavelength. The longer the wavelength, the lower the pitch. The 'height' of the wave is its amplitude. The amplitude determines how loud a sound will be. Greater amplitude means the sound will be louder.

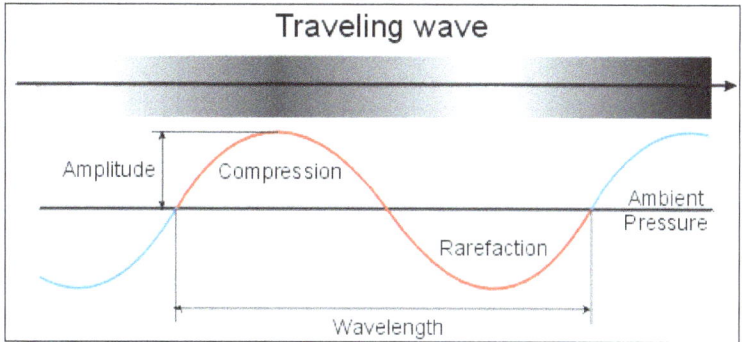

Interference

When two waves meet, there can be two kinds of interference patterns; constructive and destructive. Constructive inteference is when two waveforms are added together. The peaks add with the peaks, and the troughs add with the troughs, creating a louder sound. Destructive interference occurs when two waves are out of phase (the peaks on one line up with troughs on the other). In this, the peaks cancel out the troughs, creating a diminished waveform. For example, if two waveforms that are exactly the same are added, the amplitude doubles, but when two opposite waveforms are added, they cancel out, leaving silence.

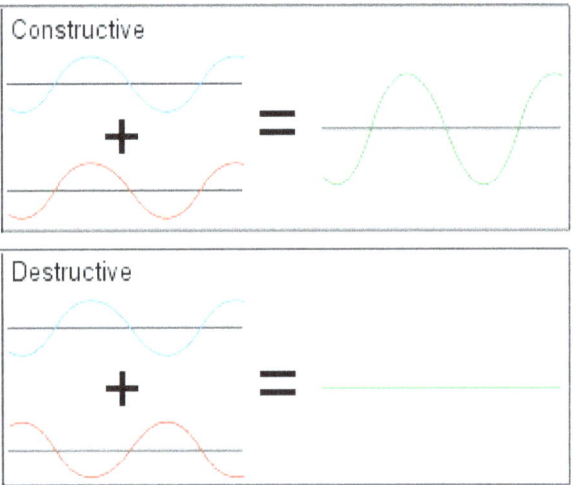

Standing Waves

Vibration inside a tube forms a standing wave. A standing wave is the result of the wave reflecting off the end of the tube (whether closed or open) and interfering with itself. When sound is produced in an instrument by blowing it, only the waves that will fit in the tube resonate, while other frequencies are lost. The longest wave that can fit in the tube is the fundamental, while other waves that fit are overtones. Overtones are multiples of the fundamental. The areas of highest vibration are called antinodes (labeled 'A' on the diagram), while the areas of least vibration are called nodes (labeled 'N' in the diagram). In an open pipe, the ends are antinodes. However, in a pipe closed at one end, the closed end is a node, while the blown end is an antinode. Thus, closed pipes yield only half the harmonics.

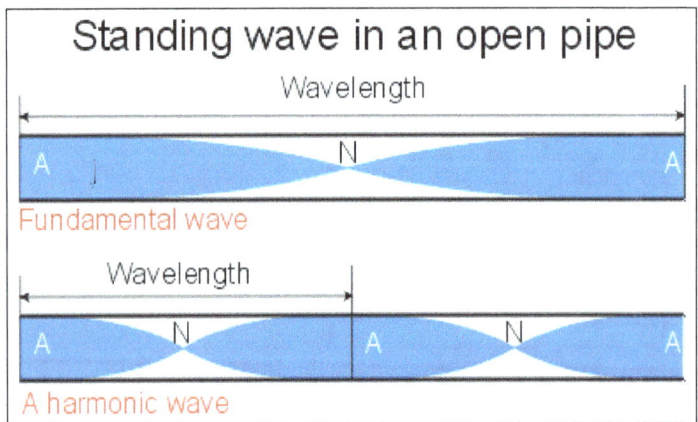

Transverse Waves

If a string that is fixed on both ends is bowed or plucked, such as in a violin, vibrations are formed that are in a standing wave pattern, having nodes at the fixed ends, and an antinode in the center. Several harmonics are also produced, in a similar way to the standing wave.

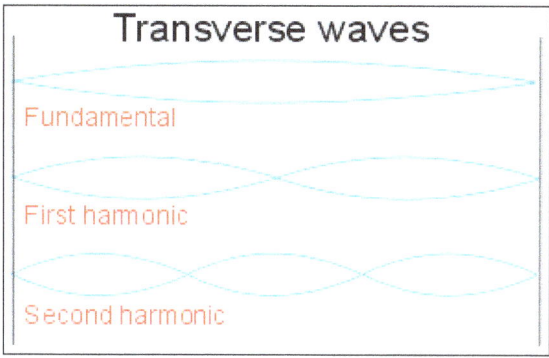

Overtones

Overtones are the other frequencies besides the fundamental that exist in musical instruments. Instruments of different shapes and actions produce different overtones. The overtones combine to form the characteristic sound of the instrument. For example, both the waves below are the same frequency, and therefore the same note. But their overtones are different, and therefore their sounds are different. Note that the violin's jagged waveform produces a sharper sound, while the smooth waveform of the piano produces a purer sound, closer to a sine wave. Click on each wave to hear what it sounds like. Keep in mind that all are playing the same note.

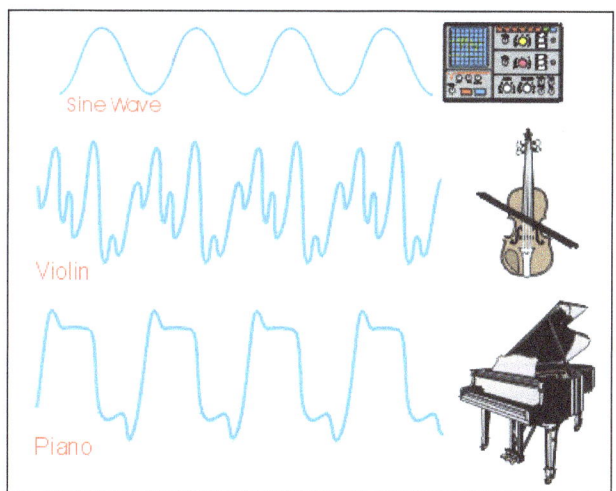

ECHO

In audio signal processing and acoustics, echo is a reflection of sound that arrives at the listener with a delay after the direct sound. The delay is directly proportional to the distance of the reflecting surface from the source and the listener. Typical examples are the echo produced by the bottom of a well, by a building, or by the walls of an enclosed room and an empty room. A true echo is a single reflection of the sound source.

Some animals use echo for location sensing and navigation, such as cetaceans (dolphins and whales) and bats.

Acoustic Phenomenon

Acoustic waves are reflected by walls or other hard surfaces, such as mountains and privacy fences. The reason of reflection may be explained as a discontinuity in the propagation medium. This can be heard when the reflection returns with sufficient magnitude and delay to be perceived distinctly. When sound, or the echo itself, is reflected multiple times from multiple surfaces, the echo is characterized as a reverberation.

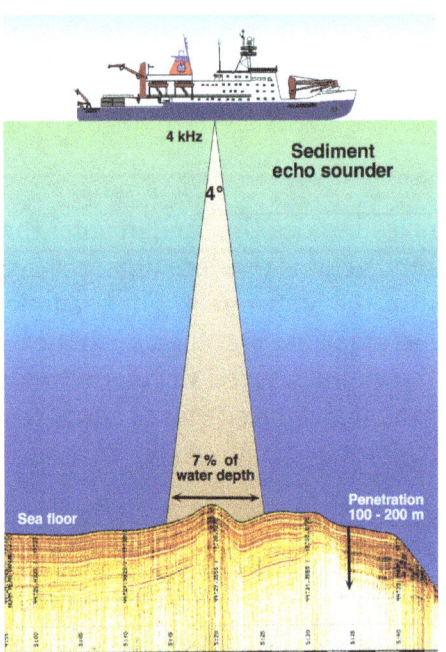

This illustration depicts the principle of sediment echo sounding, which uses a narrow beam of high energy and low frequency.

The human ear cannot distinguish echo from the original direct sound if the delay is less than 1/10 of a second. The velocity of sound in dry air is approximately 343 m/s at a temperature of 25 °C. Therefore, the reflecting object must be more than 17.2m from the sound source for echo to be perceived by a person located at the source. When a sound produces an echo in two seconds, the reflecting object is 343m away. In nature, canyon walls or rock cliffs facing water are the most common natural settings for hearing echoes. The strength of echo is frequently measured in dB sound pressure level (SPL) relative to the directly transmitted wave. Echoes may be desirable (as in sonar) or undesirable (as in telephone systems).

5

Fundamental Theories of Physics

Ample number of theories are used to study the concepts and applications related to physics. Fundamental theories of physics include Quantum field theory, M-theory, effective field theory, lattice field theory, etc. This chapter discusses these fundamental theories related to physics in detail.

QUANTUM FIELD THEORY

Quantum Field Theory (QFT) is the mathematical and conceptual framework for contemporary elementary particle physics. In a rather informal sense QFT is the extension of quantum mechanics (QM), dealing with particles, over to fields, i.e. systems with an infinite number of degrees of freedom. In the last few years QFT has become a more widely discussed topic in philosophy of science, with questions ranging from methodology and semantics to ontology. QFT taken seriously in its metaphysical implications seems to give a picture of the world which is at variance with central classical conceptions of particles and fields, and even with some features of QM.

The following sketches how QFT describes fundamental physics and what the status of QFT is among other theories of physics. Since there is a strong emphasis on those aspects of the theory that are particularly important for interpretive inquiries, it does not replace an introduction to QFT as such. One main group of target readers are philosophers who want to get a first impression of some issues that may be of interest for their own work, another target group are physicists who are interested in a philosophical view upon QFT.

In contrast to many other physical theories there is no canonical definition of what QFT is. Instead one can formulate a number of totally different explications, all of which have their merits and limits. One reason for this diversity is the fact that QFT has grown successively in a very complex way. Another reason is that the interpretation of QFT is particularly obscure, so that even the spectrum of options is not clear. Possibly the best and most comprehensive understanding of QFT is gained by dwelling on its relation to other physical theories, foremost with respect to QM, but also with respect to classical electrodynamics, Special Relativity Theory (SRT) and Solid State Physics or more generally Statistical Physics. However, the connection between QFT and these theories is also complex and cannot be neatly described step by step.

If one thinks of QM as the modern theory of one particle (or, perhaps, a very few particles), one can then think of QFT as an extension of QM for analysis of systems with many particles—and therefore with a large number of degrees of freedom. In this respect going from QM to QFT is not inevitable but rather beneficial for pragmatic reasons. However, a general threshold is crossed when it comes to fields, like the electromagnetic field, which are not merely difficult but impossible to deal with in the frame of QM. Thus the transition from QM to QFT allows treatment of both particles and fields within a uniform theoretical framework. (As an aside, focusing on the number of particles, or degrees of freedom respectively, explains why the famous renormalization group methods can be applied in QFT as well as in Statistical Physics. The reason is simply that both disciplines study systems with a large or an infinite number of degrees of freedom, either because one deals with fields, as does QFT, or because one studies the thermodynamic limit, a very useful artifice in Statistical Physics.). Moreover, issues regarding the number of particles under consideration yield yet another reason why we need to extend QM. Neither QM nor its immediate relativistic extension with the Klein-Gordon and Dirac equations can describe systems with a variable number of particles. However, obviously this is essential for a theory that is supposed to describe scattering processes, where particles of one kind are destroyed while others are created.

One gets a very different kind of access to what QFT is when focusing on its relation to QM and SRT. One can say that QFT results from the successful reconciliation of QM and SRT. In order to understand the initial problem one has to realize that QM is not only in a potential conflict with SRT, more exactly: the locality postulate of SRT, because of the famous EPR correlations of entangled quantum systems. There is also a manifest contradiction between QM and SRT on the level of the dynamics. The Schrödinger equation, i.e. the fundamental law for the temporal evolution of the quantum mechanical state function, cannot possibly obey the relativistic requirement that all physical laws of nature be invariant under Lorentz transformations. The Klein-Gordon and Dirac equations, resulting from the search for relativistic analogues of the Schrödinger equation in the 1920s, do respect the requirement of Lorentz invariance. Nevertheless, ultimately they are not satisfactory because they do not permit a description of fields in a principled quantum-mechanical way.

Fortunately, for various phenomena it is legitimate to neglect the postulates of SRT, namely when the relevant velocities are small in relation to the speed of light and when the kinetic energies of the particles are small compared to their mass energies mc2. And this is the reason why non-relativistic QM, although it cannot be the correct theory in the end, has its empirical successes. But it can never be the appropriate framework for electromagnetic phenomena because electrodynamics, which prominently encompasses a description of the behavior of light, is already relativistically invariant and therefore incompatible with QM. Scattering experiments are another context in which QM fails. Since the involved particles are often accelerated almost up to the speed of light, relativistic effects can no longer be neglected. For that reason scattering experiments can only be correctly grasped by QFT.

Unfortunately, the catchy characterization of QFT as the successful merging of QM and SRT has its limits. On the one hand, as already mentioned above, there also is a relativistic QM, with the Klein-Gordon- and the Dirac-equation among their most famous results. On the other hand, and this may come as a surprise, it is possible to formulate a non-relativistic version of QFT. The nature of QFT thus cannot simply be that it reconciles QM with the requirement of relativistic invariance. Consequently, for a discriminating criterion it is more appropriate to say that only QFT, and not QM, allows describing systems with an infinite number of degrees of freedom, i.e. fields (and systems in the thermodynamic limit). According to this line of reasoning, QM would be the modern (as opposed to classical) theory of particles and QFT the modern theory of particles and fields. Unfortunately however, and this shall be the last turn, even this gloss is not untarnished. There is a widely discussed no-go theorem by Malament with the following proposed interpretation: Even the quantum mechanics of one single particle can only be consonant with the locality principle of special relativity theory in the framework of a field theory, such as QFT. Hence ultimately, the characterization of QFT, on the one hand, as the quantum physical description of systems with an infinite number of degrees of freedom, and on the other hand, as the only way of reconciling QM with special relativity theory, are intimately connected with one another.

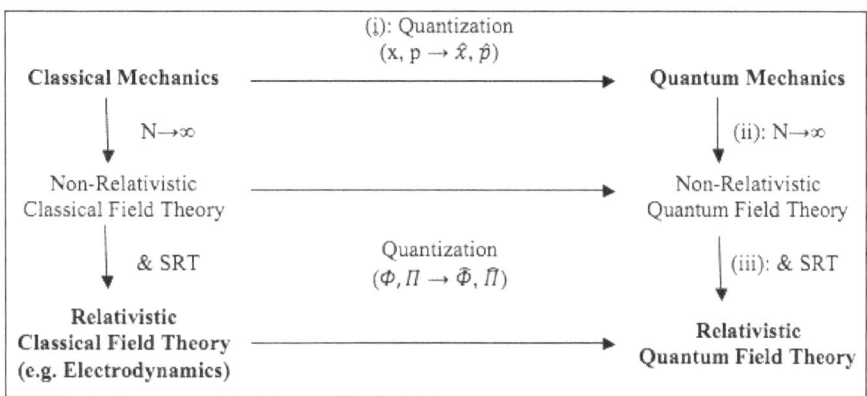

The diagram depicts the relations between different theories, where Non-Relativistic Quantum Field Theory is not a historical theory but rather an ex post construction that is illuminating for conceptual purposes. Theoretically, [(i), (ii), (iii)], [(ii), (i), (iii)] and [(ii), (iii), (i)] are three possible ways to get from Classical Mechanics to Relativistic Quantum Field Theory. But note that this is meant as a conceptual decomposition; history didn't go all these steps separately. On the one hand, by good luck, so to say, classical electrodynamics is relativistically invariant already, so that its successful quantization leads directly to Relativistic Quantum Field Theory. On the other hand, some would argue that the only way to reconcile QM and SRT is in terms of a field theory, so that (ii) and (iii) would coincide. Note that the steps (i), (ii) and (iii), i.e. quantization, transition to an infinite number of degrees of freedom, and reconciliation with SRT, are all ontologically relevant. In other words, by these steps the nature of the physical entities the theories talk about may change fundamentally.

The Basic Structure of the Conventional Formulation

The Lagrangian Formulation of QFT

The crucial step towards *quantum* field theory is in some respects analogous to the corresponding quantization in quantum mechanics, namely by imposing commutation relations, which leads to operator valued quantum fields. The starting point is the classical Lagrangian formulation of mechanics, which is a so-called analytical formulation as opposed to the standard version of Newtonian mechanics. A generalized notion of momentum (the *conjugate* or *canonical* momentum) is defined by setting $p = \partial L/\partial \dot{q}$, where L is the Lagrange function $L = T - V$ (T is the kinetic energy and V the potential) and $\dot{q} \equiv dq/dt$. This definition can be motivated by looking at the special case of a Lagrange function with a potential V which depends only on the position so that (using Cartesian coordinates) $\partial L/\partial \dot{x} = (\partial/\partial \dot{x})(m\dot{x}^2/2) = m\dot{x} = p_x$. Under these conditions the generalized momentum coincides with the usual mechanical momentum. In classical Lagrangian *field* theory one associates with the given field φ a second field, namely the conjugate field,

$$\pi = \partial L / \partial \dot{\varphi}$$

where L is a Lagrangian density. The field φ and its conjugate field π are the direct analogues of the canonical coordinate q and the generalized (canonical or conjugate) momentum p in classical mechanics of point particles.

In both cases, QM and QFT, requiring that the canonical variables satisfy certain commutation relations implies that the basic quantities become operator valued. From a physical point of view this shift implies a restriction of possible measurement values for physical quantities some (but not all) of which can have their values only in discrete steps now. In QFT the canonical commutation relations for a field φ and the corresponding conjugate field π are,

$$\left[\varphi(\mathbf{x},t),\ \pi(\mathbf{y},t)\right] = i\delta^3(\mathbf{x}-\mathbf{y})$$

$$\left[\varphi(\mathbf{x},t),\ \varphi(\mathbf{y},t)\right] = \left[\pi(\mathbf{x},t),\ \pi(\mathbf{y},t)\right] = 0$$

which are equal-time commutation relations, i.e., the commutators always refer to fields at the same time. It is not obvious that the equal-time commutation relations are Lorentz invariant but one can formulate a manifestly covariant form of the canonical commutation relations. If the field to be quantized is not a bosonic field, like the Klein-Gordon field or the electromagnetic field, but a fermionic field, like the Dirac field for electrons one has to use anticommutation relations.

While there are close analogies between quantization in QM and in QFT there are also important differences. Whereas the commutation relations in QM refer to a quantum object with three degrees of freedom, so that one has a set of 15 equations, the commutation relations in QFT do in fact comprise an infinite number of equations, namely for each of the infinitely many space-time 4-tuples $(\mathbf{x},t)$ there is a new set of commutation relations. This infinite number of degrees of freedom embodies the field character of QFT.

It is important to realize that the operator valued field $\varphi(\mathbf{x},t)$ in QFT is *not* analogous to the wavefunction $\psi(\mathbf{x},t)$ in QM, i.e., the quantum mechanical state in its position representation. While the wavefunction in QM is acted upon by observables/operators, in QFT it is the (operator valued) field itself which acts on the space of states. In a certain sense the single particle wave functions have been transformed, via their reinterpretation as operator valued quantum fields, into observables. This step is sometimes called 'second quantization' because the single particle wave equations in relativistic QM already came about by a quantization procedure, e.g., in the case of the Klein-Gordon equation by replacing position and momentum by the corresponding quantum mechanical operators. Afterwards the solutions to these single particle wave equations, which are states in relativistic QM, are considered as classical fields, which can be subjected to the canonical quantization procedure of QFT. The term 'second quantization' has often been criticized partly because it blurs the important fact that the single particle wave function φ in relativistic QM and the operator valued quantum field φ are fundamentally different kinds of entities despite their connection in the context of discovery.

It must be emphasized that both in QM and QFT states *and* observables are equally important. However, to some extent their roles are switched. While states in QM can have a concrete spatio-temporal meaning in terms of probabilities for position measurements, in QFT states are abstract entities and it is the quantum field operators that seem to allow for a spatio-temporal interpretation.

Interaction

Up to this point, the aim was to develop a free field theory. Doing so does not only neglect interaction with other particles (fields), it is even unrealistic for one free particle because it interacts with the field that it generates itself. For the description of interactions—such as scattering in particle colliders—we need certain extensions and modifications of the formalism. The immediate contact between scattering experiments and QFT is given by the scattering or S-matrix which contains all the relevant predictive information about, e.g., scattering cross sections. In order to calculate the S-matrix the interaction Hamiltonian is needed. The Hamiltonian can in turn be derived from the Lagrangian density by means of a Legendre transformation.

In order to discuss interactions one introduces a new representation, the *interaction picture*, which is an alternative to the Schrödinger and the Heisenberg picture. For the interaction picture one splits up the Hamiltonian, which is the generator of time-translations, into two parts $H = H_0 + H_{int}$, where H_0 describes the free system, i.e., without interaction, and gets absorbed in the definition of the fields and H_{int} is the interaction part of the Hamiltonian, or short the 'interaction Hamiltonian'. Using the interaction picture is advantageous because the equations of motion as well as, under certain conditions, the commutation relations are the same for interacting fields as for free fields. Therefore, various results that were established for free fields can still be used in the case of interacting fields. The central instrument for the description of interaction is again the S-matrix, which expresses the connection between in and out states by specifying the transition amplitudes. In QED, for instance, a state $|in\rangle$ describes one particular configuration of electrons, positrons and photons, i.e., it describes how many of these particles there are and which momenta, spins and polarizations they have before the interaction. The S-matrix supplies the probability that this state goes over to a particular $|out\rangle$ state, e.g., that a particular counter responds after the interaction. Such probabilities can be checked in experiments.

The canonical formalism of QFT is only applicable in the case of free fields since the inclusion of interaction leads to infinities. For this reason perturbation theory makes up a large part of most publications on QFT. The importance of perturbative methods is understandable realizing that they establish the immediate contact between theory and experiment. Although the techniques of perturbation theory have become ever more sophisticated it is somewhat disturbing that perturbative methods could not be avoided even in principle. One reason for this unease is that perturbation theory is felt to be rather a matter of (highly sophisticated) craftsmanship than of understanding nature. Accordingly, the corpus of perturbative methods plays a small role in the philosophical investigations of QFT. What does matter, however, is in which sense the consideration of interaction effects the general framework of QFT.

Gauge Invariance

Some theories are distinguished by being *gauge invariant*, which means that *gauge transformationsof* certain terms do not change any observable quantities. Requiring gauge invariance provides an elegant and systematic way of introducing terms for interacting fields. Moreover, gauge invariance plays an important role in selecting theories. The prime example of an intrinsically gauge invariant theory is electrodynamics. In the potential formulation of Maxwell's equations one introduces the vector potential A and the scalar potential φ, which are linked to the magnetic *field* $B(x,t)$ and the electric field $E(x,t)$ by:

$$\mathbf{B} = \nabla \times \mathbf{A}$$
$$\mathbf{E} = -(\partial \mathbf{A} / \partial t) - \nabla \varphi$$

or covariantly,

$$F^{\mu\nu} = \partial^{\mu} A^{\nu} - \partial^{\nu} A^{\mu}$$

where $F^{\mu\nu}$ is the electromagnetic field tensor and $A^{\mu} = (\varphi, A)$ the 4-vector potential. The important point in the present context is that given the identification (3.3), or (3.4), there remains a certain flexibility or freedom in the choice of A and φ, or A^{μ}. In order to see that, consider the so-called *gauge transformations,*

$$\mathbf{A} \rightarrow \mathbf{A} - \nabla \psi$$

$$\varphi \rightarrow \varphi + \partial \chi / \partial t$$

or covariantly,

$$A^{\mu} \rightarrow A^{\mu} + \partial^{\mu} \chi$$

where χ is a scalar function (of space and time or of space-time) which can be chosen arbitrarily. Inserting the transformed potentials into equations, one can see that the electric field E and the magnetic field B, or covariantly the electromagnetic field tensor $F^{\mu\nu}$, are not effected by a gauge transformation of the potentials. Since only the electric field E and the magnetic field B, and quantities constructed from them, are observable, whereas the vector potential itself is not, nothing physical seems to be changed by a gauge transformation because it leaves E and Bunaltered. Note that gauge invariance is a kind of symmetry that does not come about by space-time transformations.

In order to link the notion of gauge invariance to the Lagrangian formulation of QFT one needs a more general form of gauge transformations which applies to the field operator φ and which is supplied by,

$$\varphi \rightarrow e^{-i\Lambda} \varphi$$

$$\varphi^{*} \rightarrow e^{i\Lambda} \varphi^{*}$$

where Λ is an arbitrary real constant. Equations (3.7) describe a *global gauge transformation-* whereas a *local gauge transformation,*

$$\varphi(x) \;\rightarrow\; e^{-i\alpha(x)}\varphi(x)$$

varies with x.

It turned out that requiring invariance under local gauge transformations supplies a systematic way for finding the equations describing fundamental interactions. For instance, starting with the Lagrangian for a free electron, the requirement of local gauge invariance can only be fulfilled by introducing additional terms, namely those for the electromagnetic field. Gauge invariance can be captured by certain symmetry groups: U(1) for electromagnetic, SU(2)⊗U(1) for electroweak and SU(3) for strong interaction. This is an important basis for unification programs, as is the analogy to general relativity where a local gauge symmetry is associated with the gravitational field. Moreover, it turned out that only gauge invariant quantum field theories are renormalizable. All this can be taken to show that a mathematically rich theory, with surplus structures, can be very valuable in the construction of theories.

Auyang emphasizes the general conceptual significance of invariance principles; Redhead and Martin focus specifically on gauge symmetries. Healey and Lyre discuss the ontological significance of gauge theories, among other things concerning the Aharanov-Bohm effect and ontic structural realism.

Effective Field Theories and Renormalization

In the 1970s a program emerged in which the theories of the standard model of elementary particle physics are considered as effective field theories (EFTs) which have a common quantum field theoretical framework. EFTs describe relevant phenomena only in a certain domain since the Lagrangian contains only those terms that describe particles which are relevant for the respective range of energy. EFTs are inherently approximative and change with the range of energy considered. EFTs are only applicable on a certain energy scale, i.e., they only describe phenomena in a certain range of energy. Influences from higher energy processes contribute to average values but they cannot be described in detail. This procedure has no severe consequences since the details of low-energy theories are largely decoupled from higher energy processes. Both domains are only connected by altered coupling constants and the renormalization group describes how the coupling constants depend on the energy.

The main idea of EFTs is that theories, i.e., in particular the Lagrangians, depend on the energy of the phenomena which are analysed. The physics changes by switching to a different energy scale, e.g., new particles can be created if a certain energy threshold is exceeded. The dependence of theories on the energy scale distinguishes QFT from, e.g., Newton's theory of gravitation where the same law applies to an apple as well as to the moon. Nevertheless, laws from different energy scales are not completely independent of each other. A central aspect of considerations about this dependence are the consequences of higher energy processes on the low-energy scale.

On this background a new attitude towards renormalization developed in the 1970s, which revitalizes earlier ideas that divergences result from neglecting unknown processes of higher energies. Low-energy behavior is thus affected by higher energy processes. Since higher energies correspond to smaller distances this dependence is to be expected from an atomistic point of view. According to the reductionist program the dynamics of constituents on the microlevel should determine processes on the macrolevel, i.e., here the low-energy processes. However, as, for instance hydrodynamics shows, in practice theories from different levels are not quite as closely connected because a law which is applicable on the macrolevel can be largely independent of microlevel details. For this reason analogies with statistical mechanics play an important role in the discussion about EFTs. The basic idea of this new story about renormalization is that the influences of higher energy processes are localizable in a few structural properties which can be captured by an adjustment of parameters. "In this picture, the presence of infinities in quantum field theory is neither a disaster, nor an asset. It is simply a reminder of a practical limitation—we do not know what happens at distances much smaller than those we can look at directly". This new attitude supports the view that renormalization is the appropriate answer to the change of fundamental interactions when the QFT is applied to processes on different energy scales. The price one has to pay is that EFTs are only valid in a limited domain and should be considered as approximations to better theories on higher energy scales. This prompts the important question whether there is a last fundamental theory in this tower of EFTs which supersede each other with rising energies. Some people conjecture that this deeper theory could be a string theory, i.e., a theory which is not a field theory any more. Or should one ultimately expect from physics theories that they are only valid as approximations and in a limited domain? Hartmann and Castellani discuss the fate of reductionism vis-à-vis EFTs. Wallace and Fraser discuss what the successful application of renormalization methods in quantum statistical mechanics means for their role in QFT, reaching very different conclusions.

Beyond the Standard Model

The "standard model of elementary particle physics" is sometimes used almost synonymously with QFT. However, there is a crucial difference. While the standard model is a theory with a fixed ontology (understood in a prephilosophical sense), i.e. three fundamental forces and a certain number of elementary particles, QFT is rather a frame, the applicability of which is open. Thus while quantum chromodynamics (or 'QED') is a *part* of the standard model, it is an *instance* of a quantum field theory, or short "*a* quantum field theory" and not a part of QFT. This section deals with only some particularly important proposals that go beyond the standard model, but which do not necessarily break up the basic framework of QFT.

Quantum Gravity

The standard model of particle physics covers the electromagnetic, the weak and the strong interaction. However, the fourth fundamental force in nature, gravitation, has defied quantization so far. Although numerous attempts have been made in the last 80 years, and in particular very recently, there is no commonly accepted solution up to the present day. One basic problem is that the mass, length and time scales quantum gravity theories are dealing with are so extremely small that it is almost impossible to test the different proposals.

The most important extant versions of quantum gravity theories are canonical quantum gravity, loop theory and string theory. Canonical quantum gravity approaches leave the basic structure of QFT untouched and *just* extend the realm of QFT by quantizing gravity. Other approaches try to reconcile quantum theory and general relativity theory not by supplementing the reach of QFT but rather by changing QFT itself. String theory, for instance, proposes a completely new view concerning the most fundamental building blocks: It does not merely incorporate gravitation but it formulates a new theory that describes all four interactions in a unified way, namely in terms of strings.

While quantum gravity theories are very complicated and even more remote from classical thinking than QM, SRT and GRT, it is not so difficult to see why gravitation is far more difficult to deal with than the other three forces. Electromagnetic, weak and strong force all act in a given space-time. In contrast, gravitation is, according to GRT, not an interaction that takes place *in* time, but gravitational forces are identified with the curvature of space-time itself. Thus quantizing gravitation could amount to quantizing space-time, and it is not at all clear what that could mean. One controversial proposal is to deprive space-time of its fundamental status by showing how it "emerges " in some non-spatio-temporal theory. The "emergence" of space-time then means that there are certain derived terms in the new theory that have some formal features commonly associated with space-time. Rickles for an accessible and conceptually reflected introduction to quantum gravity and Wüthrich for a philosophical evaluation of the alleged need to quantize the gravitational field.

String Theory

String theory is one of the most promising candidates for bridging the gap between QFT and general relativity theory by supplying a unified theory of all natural forces, including gravitation. The basic idea of string theory is not to take particles as fundamental objects but strings that are very small but extended in one dimension. This assumption has the pivotal consequence that strings interact on an extended distance and not at a point. This difference between string theory and standard QFT is essential because it is the reason why string theory also encompasses the gravitational force which is very difficult to deal with in the framework of QFT.

It is so hard to reconcile gravitation with QFT because the typical length scale of the gravitational force is very small, namely at Planck scale, so that the quantum field theoretical assumption of point-like interaction leads to untreatable infinities. To put it another way, gravitation becomes significant (in particular in comparison to strong interaction) exactly where QFT is most severely endangered by infinite quantities. The extended interaction of strings brings it about that such infinities can be avoided. In contrast to the entities in standard quantum physics strings are not characterized by quantum numbers but only by their geometrical and dynamical properties. Nevertheless, "macroscopically" strings look like quantum particles with quantum numbers. A basic geometrical distinction is the one between open strings, i.e., strings with two ends, and closed strings which are like bracelets. The central dynamical property of strings is their mode of excitation, i.e., how they vibrate.

Reservations about string theory are mostly due to the lack of testability since it seems that there are no empirical consequences which could be tested by the methods which are, at least up to now, available to us. The reason for this "problem" is that the length scale of strings is in the average the same as the one of quantum gravity, namely the Planck length of approximately

10^{-33} centimeters which lies far beyond the accessibility of feasible particle experiments. But there are also other peculiar features of string theory which might be hard to swallow. One of them is the fact that string theory implies that space-time has 10, 11 or even 26 dimensions. In order to explain the appearance of only four space-time dimensions string theory assumes that the other dimensions are somehow folded away or "compactified" so that they are no longer visible. An intuitive idea can be gained by thinking of a macaroni which is a tube, i.e., a two-dimensional piece of pasta rolled together, but which looks from the distance like a one-dimensional string.

Despite of the problems of string theory, physicists do not abandon this project, partly because many think that, among the numerous alternative proposals for reconciling quantum physics and general relativity theory, string theory is still the best candidate, with "loop quantum gravity" as its strongest rival. Correspondingly, string theory has also received some attention within the philosophy of physics community in recent years. Probably the first philosophical investigation of string theory is Weingard in Callender & Huggett, an anthology with further related articles. Dawid argues that string theory has significant consequences for the philosophical debate about realism, namely that it speaks against the plausibility of anti-realistic positions. Johansson and Matsubara assess string theory from various different methodological perspectives, reaching conclusions in disagreement with Dawid . Standard introductory monographs on string theory are Polchinski and Kaku . Greene is a very successful popular introduction. An interactive website with a nice elementary introduction is 'Stringtheory.com'.

Axiomatic Reformulations of QFT

Deficiencies of the Conventional Formulation of QFT

From the 1930s onwards the problem of infinities as well as the potentially heuristic status of the Lagrangian formulation of QFT stimulated the search for reformulations in a concise and eventually axiomatic manner. A number of further aspects intensified the unease about the standard formulation of QFT. The first one is that quantities like total charge, total energy or total momentum of a field are unobservable since their measurement would have to take place in the whole universe. Accordingly, quantities which refer to infinitely extended regions of space-time should not appear among the observables of the theory as they do in the standard formulation of QFT. Another problematic feature of standard QFT is the idea that QFT is about field values at points of space-time. The mathematical aspect of the problem is that a field at a point, $\varphi(x)$, is not an operator in a Hilbert space. The physical counterpart of the problem is that it would require an infinite amount of energy to measure a field at a point of space-time. One way to handle this situation—and one of the starting points for axiomatic reformulations of QFT—is not to consider fields at a point but instead fields which are smeared out in the vicinity of that point using certain functions, so-called test functions. The result is a smeared field $\varphi(f) = \int \varphi(x)f(x)dx$ with supp$(f) \subset$ O, where supp(f) is the support of the test function f and O is a bounded open region in Minkowski space-time.

The third important problem for standard QFT which prompted reformulations is the existence of inequivalent representations. In the context of quantum mechanics, Schrödinger, Dirac, Jordan and von Neumann realized that Heisenberg's matrix mechanics and Schrödinger's wave mechanics are just two (unitarily) equivalent representations of the same underlying abstract structure, i.e., an abstract Hilbert space H and linear operators acting on this space. In other words, we are merely dealing with two different ways for representing the same physical reality, and it is possible to switch between these different representations by means of a unitary transformation, i.e. an operation that is analogous to an innocuous rotation of the frame of reference. *Representations* of some given algebra or group are sets of mathematical objects, like numbers, rotations or more abstract transformations (e.g. differential operators) together with a binary operation (e.g. addition or multiplication) that combines any two elements of the algebra or group, such that the structure of the algebra or group to be represented is preserved. This means that the combination of any two elements in the representation space, say a and b, leads to a third element which corresponds to the element that results when you combine the elements corresponding to a and b in the algebra or group that is represented. In 1931 von Neumann gave a detailed proof (of a conjecture by Stone) that the canonical commutation relations (CCRs) for position coordinates and their conjugate momentum coordinates in configuration space fix the representation of these two sets of operators in Hilbert space up to unitary equivalence (von Neumann's uniqueness theorem). This means that the specification of the purely algebraic CCRs suffices to describe a particular physical system.

In quantum *field* theory, however, von Neumann's uniqueness theorem looses its validity since here one is dealing with an infinite number of degrees of freedom. Now one is confronted with a multitude of *inequivalent* irreducible representations of the CCRs and it is not obvious what this means physically and how one should cope with it. Since the troublesome inequivalent representations of the CCRs that arise in QFT are all *irreducible* their inequivalence is not due to the fact that some are reducible while others are not (a representation is *reducible* if there is an invariant subrepresentation, i.e. a subset which alone represent the CCRs already). Since inequivalent irreducible representations (short: IIRs) seem to describe different physical states of affairs it is no longer legitimate to simply choose the most convenient representation, just like choosing the most convenient frame of reference. The acuteness of this problem is not immediately clear, since prima facie it is possibly that all but one of the IIRs are physically irrelevant, i.e. mathematical artefacts of a redundant formalism. However, although apparently this applies to most of the available IIRs, it seems that a number of irreducible representations of the CCRs remain that are inequivalent *and* physically relevant.

Algebraic Approaches to QFT

According to the algebraic point of view *algebras* of observables rather than observables themselves in a particular representation should be taken as the basic entities in the mathematical description of quantum physics; thereby avoiding the above-mentioned problems from the outset. In standard QM the algebraic point of view in terms of C^*-algebras makes no notable difference to the usual Hilbert space formulation since both formalisms are equivalent. However, in QFT this is no longer the case since the infinite number of degrees of freedom leads to unitarily *inequivalent* irreducible representations of a C^*-algebra. Thus sticking to the usual Hilbert space formulation tacitly implies choosing one particular representation. The notion of C^*-algebras, introduced abstractly by Gelfand and Neumark in 1943 and named this way by Segal in 1947, generalizes the notion of the algebra B(H) of all bounded operators on a Hilbert space H, which is also the most important example for a C^*-algebra. In fact, it can be shown that any C^*-algebra is isomorphic to a (norm-closed, self-adjoint) algebra of bounded operators on a Hilbert space. The boundedness (and self-adjointness) of the operators is the reason why C^*-algebras are considered as ideal for representing physical observables. The 'C' indicates that one is dealing with a complex vector space and the '*' refers to the operation that maps an element A of an algebra to its *involution* (or adjoint) A^*, which generalizes the conjugate complex of complex numbers to operators. This involution is needed in order to define the crucial norm property of C^*-algebras, which is of central importance for the proof of the above isomorphism claim.

Another point where algebraic formulations are advantageous derives from the fact that two quantum fields are physically equivalent when they generate the same algebras of local observables. Such equivalent quantum field theories belong to the same so-called Borchers class which entails that they lead to the same S-matrix. As Haag stresses, fields are only an instrument in order to "coordinatize" observables, more precisely: sets of observables, with respect to different finite spacetime regions. The choice of a particular field system is to a certain degree conventional, namely as long as it belongs to the same Borchers class. Thus it is more appropriate to consider these algebras, rather than quantum fields, as the fundamental entities in QFT.

A prominent attempt to axiomatise QFT is Wightman's field axiomatics from the early 1950s. Wightman imposed axioms on polynomial algebras P(O) of smeared fields, i.e., sums of products of smeared fields in finite space-time regions O. A crucial point of this approach is replacing the mapping $x \rightarrow \varphi(x)$ by O $\rightarrow$ P(O). While the usage of unbounded field operators makes Wightman's approach mathematically cumbersome, Algebraic Quantum Field Theory (AQFT)—arguably the most successful attempt to reformulate QFT axiomatically—employs only bounded operators. AQFT originated in the late 1950s by the work of Haag and quickly advanced in collaboration with Araki and Kastler. AQFT itself exists in two versions, concrete AQFT (Haag-Araki) and abstract AQFT. The concrete approach uses von Neumann algebras (or W^*-algebras), the abstract one C^*-algebras. The adjective 'abstract' refers to the fact that in this approach the algebras are characterized in an abstract fashion and not by explicitly using operators on a Hilbert space. In standard QFT, the CCRs together with the field equations can be used for the same purpose, i.e., an abstract characterization. One common aim of these axiomatizations of QFT is avoiding the usual approximations of standard QFT. However, trying to do this in a strictly axiomatic way, one only gets 'reformulations' which are not as rich as standard QFT. As Haag concedes, the "algebraic approach has given us a frame and a language not a theory".

Basic Ideas of AQFT

One of the crucial ideas of AQFT is taking so-called *nets of algebras* as basic for the mathematical description of a quantum physical system. A decade earlier, Segal used a single C^*-algebra—generated by all bounded operators—and dismissed the availability of inequivalent representations as irrelevant to physics. Against this approach Haag argued that inequivalent representations can be understood physically by realizing that the important physical information in a quantum field theory is not contained in individual algebras but in the net of algebras, i.e. in the mapping O $\rightarrow$ A(O) from finite space-time regions to algebras of local observables. The crucial point is that it is *not* necessary to specify observables explicitly in order to fix physically meaningful quantities. The very way how algebras of local observables are linked to space-time regions is sufficient to supply observables with physical significance. It is the partition of the algebra A_{loc} of *all* local observables into subalgebras which contains physical information about the observables, i.e., it is the net structure of algebras which matters.

Physically the most important notion of AQFT is the principle of *locality* which has an external as well as an internal aspect. The external aspect is the fact that AQFT considers only observables connected with finite regions of space-time and not global observables like the total charge or the total energy momentum vector which refer to infinite space-time regions. This approach was motivated by the operationalistic view that QFT is a statistical theory about local measurement outcomes with all the experimental information coming from measurements in finite space-time regions. Accordingly everything is expressed in terms of *local algebras* of observables. The internal aspect of locality is that there is a constraint on the observables of such local algebras: All observables of a local algebra connected with a space-time region O are required to commute with all observables of another algebra which is associated with a space-time region O' that is space-like separated from O. This principle of (Einstein) *causality* is the main relativistic ingredient of AQFT.

The basic structure upon which the assumptions or conditions of AQFT are imposed are local observables, i.e., self-adjoint elements in local (non-commutative) von Neumann-algebras, and physical states, which are identified as positive, linear, normalized functionals which map elements of local algebras to real numbers. States can thus be understood as assignments of expectation values to observables. One can group the assumptions of AQFT into relativistic axioms, such as locality and covariance, general physical assumptions, like isotony and spectrum condition, and finally technical assumptions which are closely related to the mathematical formulation.

As a reformulation of QFT, AQFT is expected to reproduce the main phenomena of QFT, in particular properties which are characteristic of it being a field theory, like the existence of antiparticles, internal quantum numbers, the relation of spin and statistics, etc. That this aim could not be achieved on a purely axiomatic basis is partly due to the fact that the connection between the respective key concepts of AQFT and QFT, i.e., observables and quantum fields, is not sufficiently clear. It turned out that the main link between observable algebras and quantum fields are *superselection rules*, which put restrictions on the set of all observables and allow for classification schemes in terms of permanent or essential properties.

Introductions to AQFT are provided by the monographs Haag and Horuzhy as well as the overview articles Haag & Kastler, Roberts and Buchholz. Streater & Wightman is an early pioneering monograph on axiomatic QFT. Bratteli & Robinson emphasize mathematical aspects.

AQFT and the Philosopher

In recent years, QFT has received a lot of attention in the philosophy of physics. Most philosophers who engage in that debate rest their considerations on AQFT; for instance, Baker & Halvorson , Earman & Fraser, Fraser Halvorson & Müger, Kronz & Lupher, Kuhlmann , Lupher , Rédei & Valente and Ruetsche. While most philosophers of physics who are skeptical about this approach remained largely silent, Wallace launched an eloquent attack on the predominance of AQFT for foundational studies about QFT. To be sure, Wallace emphasizes, his critique is not directed against the use of algebraic methods, e.g. when studying inequivalent representations. Rather, he aims at AQFT as a physical theory, regarded as a rival to conventional QFT (CQFT). In his evaluation, viewed from the 21st century, one has to state that CQFT succeeded, while AQFT failed, so that "to be lured away from the Standard Model by [AQFT] is sheer madness". So what may justify this drastic conclusion? On the one hand, Wallace points out that, the problem of ultraviolet divergences, which initiated the search for alternative approaches in the 1950s, was eventually solved in CQFT via the renormalization group techniques. On the other hand, AQFT never succeeded in finding realistic interacting quantum field theories in four dimensions (such as QED) that fit into their framework.

Fraser is most actively engaged in defending AQFT against Wallace's assault. She argues that consistency plays a central role in choosing between different formulations of QFT since they do not differ in their respective empirical success and AQFT fares better in this respect. Moreover, Fraser questions Wallace's crucial point in defense of CQFT, namely that the empirically successful application of renormalization group techniques in QFT removes all doubts about CQFT: The fact that renormalization in condensed matter physics and QFT are formally similar does not license Wallace's claim that there are also physical similarities concerning the freezing out of degrees of freedom at very small length scales. And if that physical analogy cannot be sustained, then the empirical success of renormalization in CQFT leaves the physical reasons for this success in the dark, in contrast to the case of condensed matter physics, where the physical basis for the empirical success of renormalization is intelligible, namely the fact that matter is discrete at atomic length scales. As a consequence, despite of the formal analogy with renormalization in condensed matter physics the empirical success of renormalization in CQFT does not, as Wallace claims, discredit the idea to work with arbitrarily small regions of spacetime, as it is done in AQFT.

Kuhlmann also advocates AQFT as the prime object for foundational studies, focusing on ontological considerations. He argues that for matters of ontology AQFT is to be preferred over CQFT because, like ontology itself, AQFT strives for a clear separation of fundamental and derived entities and a parsimonious selection of basic assumptions. CQFT, on the other hand is a grown formalism that is very good for calculations but obscures foundational issues. Moreover, Kuhlmann contends that AQFT and CQFT should not be regarded as rival research programs. Nowadays at the very least, AQFT is not meant to replace CQFT, despite of the "kill it or cure it" slogan. AQFT is suited and designed to illuminate the basic structure of QFT, but it is not and never will be the appropriate framework for the working physicist.

EFFECTIVE FIELD THEORY

In physics, an effective field theory is a type of approximation, or effective theory, for an underlying physical theory, such as a quantum field theory or a statistical mechanics model. An effective field theory includes the appropriate degrees of freedom to describe physical phenomena occurring at a chosen length scale or energy scale, while ignoring substructure and degrees of freedom at shorter distances (or, equivalently, at higher energies). Intuitively, one averages over the behavior of the underlying theory at shorter length scales to derive what is hoped to be a simplified model at longer length scales. Effective field theories typically work best when there is a large separation between length scale of interest and the length scale of the underlying dynamics. Effective field theories have found use in particle physics, statistical mechanics, condensed matter physics, general relativity, and hydrodynamics. They simplify calculations, and allow treatment of dissipation and radiation effects.

The Renormalization Group

Presently, effective field theories are discussed in the context of the renormalization group (RG) where the process of *integrating out* short distance degrees of freedom is made systematic. Although this method is not sufficiently concrete to allow the actual construction of effective field theories, the gross understanding of their usefulness becomes clear through an RG analysis. This method also lends credence to the main technique of constructing effective field theories, through the analysis of symmetries. If there is a single mass scale M in the microscopic theory, then the effective field theory can be seen as an expansion in 1/M. The construction of an effective field theory accurate to some power of 1/M requires a new set of free parameters at each order of the expansion in 1/M. This technique is useful for scattering or other processes where the maximum momentum scale k satisfies the condition k/M≪1. Since effective field theories are not valid at small length scales, they need not be renormalizable. Indeed, the ever expanding number of parameters at each order in **1/M** required for an effective field theory means that they are generally not renormalizable in the same sense as quantum electrodynamics which requires only the renormalization of two parameters.

Examples of Effective Field Theories

Fermi Theory of Beta Decay

The best-known example of an effective field theory is the Fermi theory of beta decay. This theory was developed during the early study of weak decays of nuclei when only the hadrons and leptons undergoing weak decay were known. The typical reactions studied were:

$$n \rightarrow p + e^- + \bar{\nu}_e$$
$$\mu^- \rightarrow e^- + \bar{\nu}_e + \nu_\mu.$$

This theory posited a pointlike interaction between the four fermions involved in these reactions. The theory had great phenomenological success and was eventually understood to arise from the gauge theory of electroweak interactions, which forms a part of the standard model of particle physics. In this more fundamental theory, the interactions are mediated by a flavour-changing gauge boson, the $W^\pm$. The immense success of the Fermi theory was because the W particle has mass of about 80 GeV, whereas the early experiments were all done at an energy scale of less than 10 MeV. Such a separation of scales, by over 3 orders of magnitude, has not been met in any other situation as yet.

BCS Theory of Superconductivity

Another famous example is the BCS theory of superconductivity. Here the underlying theory is of electrons in a metal interacting with lattice vibrations called phonons. The phonons cause attractive interactions between some electrons, causing them to form Cooper pairs. The length scale of these pairs is much larger than the wavelength of phonons, making it possible to neglect the dynamics of phonons and construct a theory in which two electrons effectively interact at a point. This theory has had remarkable success in describing and predicting the results of experiments on superconductivity.

Effective Field Theories in Gravity

General relativity itself is expected to be the low energy effective field theory of a full theory of quantum gravity, such as string theory or Loop Quantum Gravity. The expansion scale is the Planck mass. Effective field theories have also been used to simplify problems in General Relativity, in particular in calculating the gravitational wave signature of inspiralling finite-sized objects. The most common EFT in GR is "Non-Relativistic General Relativity" (NRGR), which is similar to the post-Newtonian expansion. Another common GR EFT is the Extreme Mass Ratio (EMR), which in the context of the inspiralling problem is called EMRI.

Other Examples

Presently, effective field theories are written for many situations.

- One major branch of nuclear physics is quantum hadrodynamics, where the interactions of hadrons are treated as a field theory, which should be derivable from the underlying theory of quantum chromodynamics. Quantum hadrodynamics is the theory of the nuclear force, similarly to quantum chromodynamics being the theory of the strong interaction and quantum electrodynamics being the theory of the electromagnetic force. Due to the smaller separation of length scales here, this effective theory has some classificatory power, but not the spectacular success of the Fermi theory.

- In particle physics the effective field theory of QCD called chiral perturbation theory has had better success. This theory deals with the interactions of hadrons with pions or kaons, which are the Goldstone bosons of spontaneous chiral symmetry breaking. The expansion parameter is the pion energy/momentum.

- For hadrons containing one heavy quark (such as the bottom or charm), an effective field theory which expands in powers of the quark mass, called the heavy quark effective theory (HQET), has been found useful.

- For hadrons containing two heavy quarks, an effective field theory which expands in powers of the relative velocity of the heavy quarks, called non-relativistic QCD (NRQCD), has been found useful, especially when used in conjunctions with lattice QCD.

- For hadron reactions with light energetic (collinear) particles, the interactions with low-energetic (soft) degrees of freedom are described by the soft-collinear effective theory (SCET).

- Much of condensed matter physics consists of writing effective field theories for the particular property of matter being studied.

- Hydrodynamics can also be treated using Effective Field Theories.

M-THEORY

M-theory is a theory in physics that unifies all consistent versions of superstring theory. The existence of such a theory was first conjectured by Edward Witten at a string theory conference at the

University of Southern California in the Spring of 1995. Witten's announcement initiated a flurry of research activity known as the second superstring revolution.

Prior to Witten's announcement, string theorists had identified five versions of superstring theory. Although these theories appeared, at first, to be very different, work by several physicists showed that the theories were related in intricate and nontrivial ways. In particular, physicists found that apparently distinct theories could be unified by mathematical transformations called S-duality and T-duality. Witten's conjecture was based in part on the existence of these dualities and in part on the relationship of the string theories to a field theory called eleven-dimensional supergravity.

Although a complete formulation of M-theory is not known, the theory should describe two- and five-dimensional objects called branes and should be approximated by eleven-dimensional supergravity at low energies. Modern attempts to formulate M-theory are typically based on matrix theory or the AdS/CFT correspondence.

According to Witten, M should stand for "magic", "mystery", or "membrane" according to taste, and the true meaning of the title should be decided when a more fundamental formulation of the theory is known.

Investigations of the mathematical structure of M-theory have spawned important theoretical results in physics and mathematics. More speculatively, M-theory may provide a framework for developing a unified theory of all of the fundamental forces of nature. Attempts to connect M-theory to experiment typically focus on compactifying its extra dimensions to construct candidate models of our four-dimensional world, although so far none has been verified to give rise to physics as observed in high energy physics experiments.

Quantum Gravity and Strings

One of the deepest problems in modern physics is the problem of quantum gravity. The current understanding of gravity is based on Albert Einstein's general theory of relativity, which is formulated within the framework of classical physics. However, nongravitational forces are described within the framework of quantum mechanics, a radically different formalism for describing physical phenomena based on probability. A quantum theory of gravity is needed in order to reconcile general relativity with the principles of quantum mechanics, but difficulties arise when one attempts to apply the usual prescriptions of quantum theory to the force of gravity.

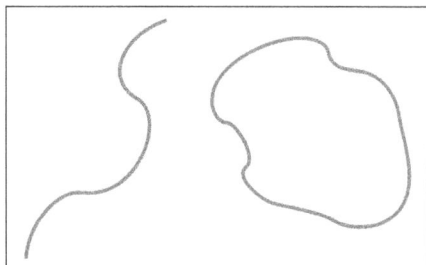

The fundamental objects of string theory are open and closed strings.

String theory is a theoretical framework that attempts to reconcile gravity and quantum mechanics. In string theory, the point-like particles of particle physics are replaced by one-dimensional objects called strings. String theory describes how strings propagate through space and interact with each other. In a given version of string theory, there is only one kind of string, which may look like a small loop or segment of ordinary string, and it can vibrate in different ways. On distance scales larger than the string scale, a string will look just like an ordinary particle, with its mass, charge, and other properties determined by the vibrational state of the string. In this way, all of the different elementary particles may be viewed as vibrating strings. One of the vibrational states of a string gives rise to the graviton, a quantum mechanical particle that carries gravitational force.

There are several versions of string theory: type I, type IIA, type IIB, and two flavors of heterotic string theory ($SO(32)$ and $E_8 \times E_8$). The different theories allow different types of strings, and the particles that arise at low energies exhibit different symmetries. For example, the type I theory includes both open strings (which are segments with endpoints) and closed strings (which form closed loops), while types IIA and IIB include only closed strings. Each of these five string theories arises as a special limiting case of M-theory. This theory, like its string theory predecessors, is an example of a quantum theory of gravity. It describes a force just like the familiar gravitational force subject to the rules of quantum mechanics.

Number of Dimensions

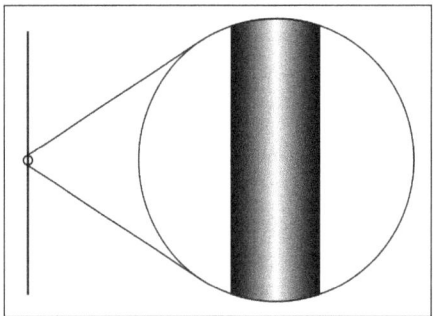

An example of compactification: At large distances, a two dimensional surface
with one circular dimension looks one-dimensional.

In everyday life, there are three familiar dimensions of space: height, width and depth. Einstein's general theory of relativity treats time as a dimension on par with the three spatial dimensions; in general relativity, space and time are not modeled as separate entities but are instead unified to a four-dimensional spacetime, three spatial dimensions and one time dimension. In this framework, the phenomenon of gravity is viewed as a consequence of the geometry of spacetime.

In spite of the fact that the universe is well described by four-dimensional spacetime, there are several reasons why physicists consider theories in other dimensions. In some cases, by modeling spacetime in a different number of dimensions, a theory becomes more mathematically tractable, and one can perform calculations and gain general insights more easily. There are also situations where theories in two or three spacetime dimensions are useful for describing phenomena in condensed matter physics. Finally, there exist scenarios in which there could actually be more than four dimensions of spacetime which have nonetheless managed to escape detection.

One notable feature of string theory and M-theory is that these theories require extra dimensions of spacetime for their mathematical consistency. In string theory, spacetime is *ten-dimensional* (nine spatial dimensions, and one time dimension), while in M-theory it is *eleven-dimensional* (ten spatial dimensions, and one time dimension). In order to describe real physical phenomena using these theories, one must therefore imagine scenarios in which these extra dimensions would not be observed in experiments.

Compactification is one way of modifying the number of dimensions in a physical theory. In compactification, some of the extra dimensions are assumed to "close up" on themselves to form circles. In the limit where these curled up dimensions become very small, one obtains a theory in which spacetime has effectively a lower number of dimensions. A standard analogy for this is to consider a multidimensional object such as a garden hose. If the hose is viewed from a sufficient distance, it appears to have only one dimension, its length. However, as one approaches the hose, one discovers that it contains a second dimension, its circumference. Thus, an ant crawling on the surface of the hose would move in two dimensions.

Dualities

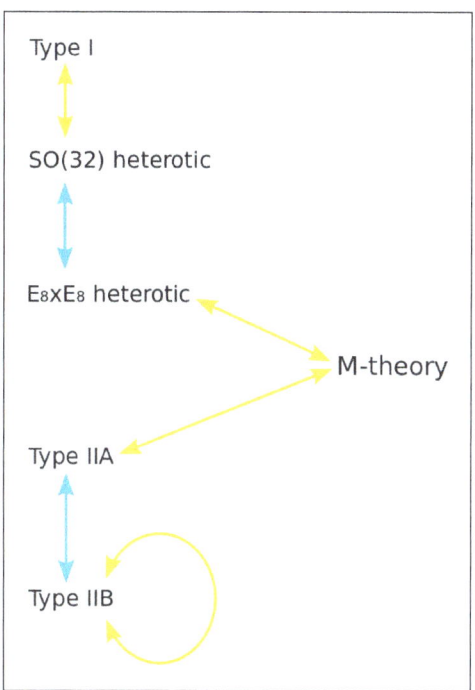

A diagram of string theory dualities. Yellow arrows indicate S-duality. Blue arrows indicate T-duality. These dualities may be combined to obtain equivalences of any of the five theories with M-theory.

Theories that arise as different limits of M-theory turn out to be related in highly nontrivial ways. One of the relationships that can exist between these different physical theories is called S-duality. This is a relationship which says that a collection of strongly interacting particles in one theory can, in some cases, be viewed as a collection of weakly interacting particles in a completely different theory. Roughly speaking, a collection of particles is said to be strongly interacting if they combine and decay often and weakly interacting if they do so infrequently. Type I string theory turns out to be equivalent by S-duality to the $SO(32)$ heterotic string theory. Similarly, type IIB string theory is related to itself in a nontrivial way by S-duality.

Another relationship between different string theories is T-duality. Here one considers strings propagating around a circular extra dimension. T-duality states that a string propagating around a circle of radius R is equivalent to a string propagating around a circle of radius $1/R$ in the sense that all observable quantities in one description are identified with quantities in the dual description. For example, a string has momentum as it propagates around a circle, and it can also wind around the circle one or more times. The number of times the string winds around a circle is called the winding number. If a string has momentum p and winding number n in one description, it will have momentum n and winding number p in the dual description. For example, type IIA string theory is equivalent to type IIB string theory via T-duality, and the two versions of heterotic string theory are also related by T-duality.

In general, the term *duality* refers to a situation where two seemingly different physical systems turn out to be equivalent in a nontrivial way. If two theories are related by a duality, it means that one theory can be transformed in some way so that it ends up looking just like the other theory. The two theories are then said to be *dual* to one another under the transformation. Put differently, the two theories are mathematically different descriptions of the same phenomena.

Supersymmetry

Another important theoretical idea that plays a role in M-theory is supersymmetry. This is a mathematical relation that exists in certain physical theories between a class of particles called bosons and a class of particles called fermions. Roughly speaking, fermions are the constituents of matter, while bosons mediate interactions between particles. In theories with supersymmetry, each boson has a counterpart which is a fermion, and vice versa. When supersymmetry is imposed as a local symmetry, one automatically obtains a quantum mechanical theory that includes gravity. Such a theory is called a supergravity theory.

A theory of strings that incorporates the idea of supersymmetry is called a superstring theory. There are several different versions of superstring theory which are all subsumed within the M-theory framework. At low energies, the superstring theories are approximated by supergravity in ten spacetime dimensions. Similarly, M-theory is approximated at low energies by supergravity in eleven dimensions.

Branes

In string theory and related theories such as supergravity theories, a brane is a physical object that generalizes the notion of a point particle to higher dimensions. For example, a point particle can be viewed as a brane of dimension zero, while a string can be viewed as a brane of dimension one. It is also possible to consider higher-dimensional branes. In dimension p, these are called p-branes. Branes are dynamical objects which can propagate through spacetime according to the rules of quantum mechanics. They can have mass and other attributes such as charge. A p-brane sweeps out a $(p+1)$-dimensional volume in spacetime called its *worldvolume*. Physicists often study fields analogous to the electromagnetic field which live on the worldvolume of a brane. The word brane comes from the word "membrane" which refers to a two-dimensional brane.

In string theory, the fundamental objects that give rise to elementary particles are the one-dimensional strings. Although the physical phenomena described by M-theory are still poorly understood, physicists know that the theory describes two- and five-dimensional branes. Much of the current research in M-theory attempts to better understand the properties of these branes.

Development

Kaluza–Klein Theory

In the early 20th century, physicists and mathematicians including Albert Einstein and Hermann Minkowski pioneered the use of four-dimensional geometry for describing the physical world. These efforts culminated in the formulation of Einstein's general theory of relativity, which relates gravity to the geometry of four-dimensional spacetime.

The success of general relativity led to efforts to apply higher dimensional geometry to explain other forces. In 1919, work by Theodor Kaluza showed that by passing to five-dimensional space-time, one can unify gravity and electromagnetism into a single force. This idea was improved by physicist Oskar Klein, who suggested that the additional dimension proposed by Kaluza could take the form of a circle with radius around 10^{-30} cm.

The Kaluza–Klein theory and subsequent attempts by Einstein to develop unified field theory were never completely successful. In part this was because Kaluza–Klein theory predicted a particle that has never been shown to exist, and in part because it was unable to correctly predict the ratio of an electron's mass to its charge. In addition, these theories were being developed just as other physicists were beginning to discover quantum mechanics, which would ultimately prove successful in describing known forces such as electromagnetism, as well as new nuclear forces that were being discovered throughout the middle part of the century. Thus it would take almost fifty years for the idea of new dimensions to be taken seriously again.

Early work on Supergravity

Edward Witten contributed to the understanding of supergravity theories.
He introduced M-theory, sparking the second superstring revolution.

New concepts and mathematical tools provided fresh insights into general relativity, giving rise to a period in the 1960s–70s now known as the golden age of general relativity. In the mid-1970s, physicists began studying higher-dimensional theories combining general relativity with super-symmetry, the so-called supergravity theories.

General relativity does not place any limits on the possible dimensions of spacetime. Although the theory is typically formulated in four dimensions, one can write down the same equations for the gravitational field in any number of dimensions. Supergravity is more restrictive because it places an upper limit on the number of dimensions. In 1978, work by Werner Nahm showed that the maximum spacetime dimension in which one can formulate a consistent supersymmetric theory is eleven. In the same year, Eugene Cremmer, Bernard Julia, and Joel Scherk of the École Normale Supérieure showed that supergravity not only permits up to eleven dimensions but is in fact most elegant in this maximal number of dimensions.

Initially, many physicists hoped that by compactifying eleven-dimensional supergravity, it might be possible to construct realistic models of our four-dimensional world. The hope was that such models would provide a unified description of the four fundamental forces of nature: electromagnetism, the strong and weak nuclear forces, and gravity. Interest in eleven-dimensional supergravity soon waned as various flaws in this scheme were discovered. One of the problems was that the laws of physics appear to distinguish between clockwise and counterclockwise, a phenomenon known as chirality. Edward Witten and others observed this chirality property cannot be readily derived by compactifying from eleven dimensions.

In the first superstring revolution in 1984, many physicists turned to string theory as a unified theory of particle physics and quantum gravity. Unlike supergravity theory, string theory was able to accommodate the chirality of the standard model, and it provided a theory of gravity consistent with quantum effects. Another feature of string theory that many physicists were drawn to in the 1980s and 1990s was its high degree of uniqueness. In ordinary particle theories, one can consider any collection of elementary particles whose classical behavior is described by an arbitrary Lagrangian. In string theory, the possibilities are much more constrained: by the 1990s, physicists had argued that there were only five consistent supersymmetric versions of the theory.

Relationships between String Theories

Although there were only a handful of consistent superstring theories, it remained a mystery why there was not just one consistent formulation. However, as physicists began to examine string theory more closely, they realized that these theories are related in intricate and nontrivial ways.

In the late 1970s, Claus Montonen and David Olive had conjectured a special property of certain physical theories. A sharpened version of their conjecture concerns a theory called $N = 4$ supersymmetric Yang–Mills theory, which describes theoretical particles formally similar to the quarks and gluons that make up atomic nuclei. The strength with which the particles of this theory interact is measured by a number called the coupling constant. The result of Montonen and Olive, now known as Montonen–Olive duality, states that $N = 4$ supersymmetric Yang–Mills theory with coupling constant g is equivalent to the same theory with coupling constant $1/g$. In other words, a system of strongly interacting particles (large coupling constant) has an equivalent description as a system of weakly interacting particles (small coupling constant) and vice versa by spin-moment.

In the 1990s, several theorists generalized Montonen–Olive duality to the S-duality relationship, which connects different string theories. Ashoke Sen studied S-duality in the context of heterotic strings in four dimensions. Chris Hull and Paul Townsend showed that type IIB string theory with a large coupling constant is equivalent via S-duality to the same theory with small coupling constant. Theorists also found that different string theories may be related by T-duality. This duality implies that strings propagating on completely different spacetime geometries may be physically equivalent.

Membranes and Fivebranes

String theory extends ordinary particle physics by replacing zero-dimensional point particles by one-dimensional objects called strings. In the late 1980s, it was natural for theorists to attempt to formulate other extensions in which particles are replaced by two-dimensional supermembranes or by higher-dimensional objects called branes. Such objects had been considered as early as 1962 by Paul Dirac, and they were reconsidered by a small but enthusiastic group of physicists in the 1980s.

Supersymmetry severely restricts the possible number of dimensions of a brane. In 1987, Eric Bergshoeff, Ergin Sezgin, and Paul Townsend showed that eleven-dimensional supergravity includes two-dimensional branes. Intuitively, these objects look like sheets or membranes propagating through the eleven-dimensional spacetime. Shortly after this discovery, Michael Duff, Paul Howe, Takeo Inami, and Kellogg Stelle considered a particular compactification of eleven-dimensional supergravity with one of the dimensions curled up into a circle. In this setting, one can imagine the membrane wrapping around the circular dimension. If the radius of the circle is sufficiently small, then this membrane looks just like a string in ten-dimensional spacetime. In fact, Duff and his collaborators showed that this construction reproduces exactly the strings appearing in type IIA superstring theory.

In 1990, Andrew Strominger published a similar result which suggested that strongly interacting strings in ten dimensions might have an equivalent description in terms of weakly interacting five-dimensional branes. Initially, physicists were unable to prove this relationship for two important reasons. On the one hand, the Montonen–Olive duality was still unproven, and so Strominger's conjecture was even more tenuous. On the other hand, there were many technical issues related to the quantum properties of five-dimensional branes. The first of these problems was solved in 1993 when Ashoke Sen established that certain physical theories require the existence of objects with both electric and magnetic charge which were predicted by the work of Montonen and Olive.

In spite of this progress, the relationship between strings and five-dimensional branes remained conjectural because theorists were unable to quantize the branes. Starting in 1991, a team of researchers including Michael Duff, Ramzi Khuri, Jianxin Lu, and Ruben Minasian considered a special compactification of string theory in which four of the ten dimensions curl up. If one considers a five-dimensional brane wrapped around these extra dimensions, then the brane looks just like a one-dimensional string. In this way, the conjectured relationship between strings and branes was reduced to a relationship between strings and strings, and the latter could be tested using already established theoretical techniques.

Second Superstring Revolution

Speaking at the string theory conference at the University of Southern California in 1995, Edward Witten of the Institute for Advanced Study made the surprising suggestion that all five superstring theories were in fact just different limiting cases of a single theory in eleven spacetime dimensions. Witten's announcement drew together all of the previous results on S- and T-duality and the appearance of two- and five-dimensional branes in string theory. In the months following Witten's announcement, hundreds of new papers appeared on the Internet confirming that the new theory involved membranes in an important way. Today this flurry of work is known as the second superstring revolution.

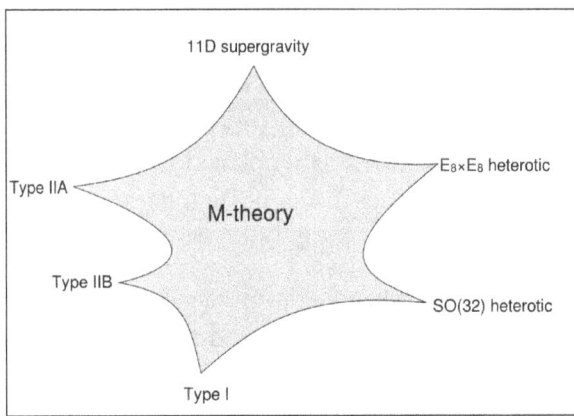

A schematic illustration of the relationship between M-theory, the five superstring theories, and eleven-dimensional supergravity. The shaded region represents a family of different physical scenarios that are possible in M-theory. In certain limiting cases corresponding to the cusps, it is natural to describe the physics using one of the six theories labeled there.

One of the important developments following Witten's announcement was Witten's work in 1996 with string theorist Petr Hořava. Witten and Hořava studied M-theory on a special spacetime geometry with two ten-dimensional boundary components. Their work shed light on the mathematical structure of M-theory and suggested possible ways of connecting M-theory to real world physics.

Origin of the Term

Initially, some physicists suggested that the new theory was a fundamental theory of membranes, but Witten was skeptical of the role of membranes in the theory. In a paper from 1996, Hořava and Witten wrote:

> As it has been proposed that the eleven-dimensional theory is a supermembrane theory but there are some reasons to doubt that interpretation, we will non-committally call it the M-theory, leaving to the future the relation of M to membranes.

In the absence of an understanding of the true meaning and structure of M-theory, Witten has suggested that the *M* should stand for "magic", "mystery", or "membrane" according to taste, and the true meaning of the title should be decided when a more fundamental formulation of the theory is known.

Matrix Theory

BFSS Matrix Model

In mathematics, a matrix is a rectangular array of numbers or other data. In physics, a matrix model is a particular kind of physical theory whose mathematical formulation involves the notion of a matrix in an important way. A matrix model describes the behavior of a set of matrices within the framework of quantum mechanics.

One important example of a matrix model is the BFSS matrix model proposed by Tom Banks, Willy Fischler, Stephen Shenker, and Leonard Susskind in 1997. This theory describes the behavior of a set of nine large matrices. In their original paper, these authors showed, among other things, that the low energy limit of this matrix model is described by eleven-dimensional supergravity. These calculations led them to propose that the BFSS matrix model is exactly equivalent to M-theory. The BFSS matrix model can therefore be used as a prototype for a correct formulation of M-theory and a tool for investigating the properties of M-theory in a relatively simple setting.

Noncommutative Geometry

In geometry, it is often useful to introduce coordinates. For example, in order to study the geometry of the Euclidean plane, one defines the coordinates x and y as the distances between any point in the plane and a pair of axes. In ordinary geometry, the coordinates of a point are numbers, so they can be multiplied, and the product of two coordinates does not depend on the order of multiplication. That is, $xy = yx$. This property of multiplication is known as the commutative law, and this relationship between geometry and the commutative algebra of coordinates is the starting point for much of modern geometry.

Noncommutative geometry is a branch of mathematics that attempts to generalize this situation. Rather than working with ordinary numbers, one considers some similar objects, such as matrices, whose multiplication does not satisfy the commutative law (that is, objects for which xy is not necessarily equal to yx). One imagines that these noncommuting objects are coordinates on some more general notion of "space" and proves theorems about these generalized spaces by exploiting the analogy with ordinary geometry.

In a paper from 1998, Alain Connes, Michael R. Douglas, and Albert Schwarz showed that some aspects of matrix models and M-theory are described by a noncommutative quantum field theory, a special kind of physical theory in which the coordinates on spacetime do not satisfy the commutativity property. This established a link between matrix models and M-theory on the one hand, and noncommutative geometry on the other hand. It quickly led to the discovery of other important links between noncommutative geometry and various physical theories.

AdS/CFT Correspondence

A tessellation of the hyperbolic plane by triangles and squares.

The application of quantum mechanics to physical objects such as the electromagnetic field, which are extended in space and time, is known as quantum field theory. In particle physics, quantum field theories form the basis for our understanding of elementary particles, which are modeled as excitations in the fundamental fields. Quantum field theories are also used throughout condensed matter physics to model particle-like objects called quasiparticles.

One approach to formulating M-theory and studying its properties is provided by the anti-de Sitter/conformal field theory (AdS/CFT) correspondence. Proposed by Juan Maldacena in late 1997, the AdS/CFT correspondence is a theoretical result which implies that M-theory is in some cases equivalent to a quantum field theory. In addition to providing insights into the mathematical structure of string and M-theory, the AdS/CFT correspondence has shed light on many aspects of quantum field theory in regimes where traditional calculational techniques are ineffective.

In the AdS/CFT correspondence, the geometry of spacetime is described in terms of a certain vacuum solution of Einstein's equation called anti-de Sitter space. In very elementary terms, anti-de Sitter space is a mathematical model of spacetime in which the notion of distance between points (the metric) is different from the notion of distance in ordinary Euclidean geometry. It is closely related to hyperbolic space, which can be viewed as a disk as illustrated on the left. This image shows a tessellation of a disk by triangles and squares. One can define the distance between points of this disk in such a way that all the triangles and squares are the same size and the circular outer boundary is infinitely far from any point in the interior.

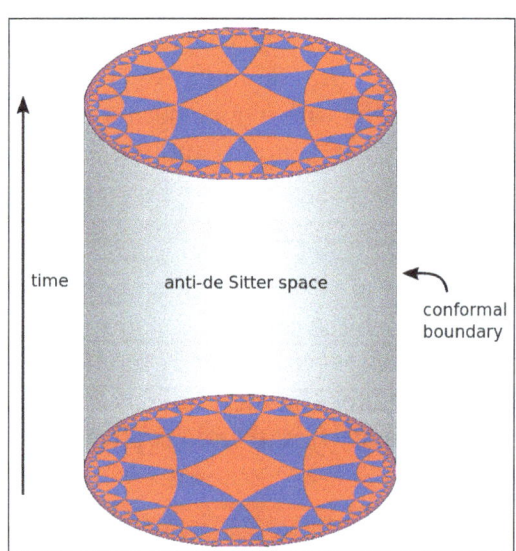

Three-dimensional anti-de Sitter space is like a stack of hyperbolic disks, each one representing the state of the universe at a given time. One can study theories of quantum gravity such as M-theory in the resulting spacetime.

Now imagine a stack of hyperbolic disks where each disk represents the state of the universe at a given time. The resulting geometric object is three-dimensional anti-de Sitter space. It looks like a solid cylinder in which any cross section is a copy of the hyperbolic disk. Time runs along the vertical direction in this picture. The surface of this cylinder plays an important role in the AdS/CFT correspondence. As with the hyperbolic plane, anti-de Sitter space is curved in such a way that any point in the interior is actually infinitely far from this boundary surface.

This construction describes a hypothetical universe with only two space dimensions and one time dimension, but it can be generalized to any number of dimensions. Indeed, hyperbolic space can have more than two dimensions and one can "stack up" copies of hyperbolic space to get higher-dimensional models of anti-de Sitter space.

An important feature of anti-de Sitter space is its boundary (which looks like a cylinder in the case of three-dimensional anti-de Sitter space). One property of this boundary is that, within a small region on the surface around any given point, it looks just like Minkowski space, the model of spacetime used in nongravitational physics. One can therefore consider an auxiliary theory in which "spacetime" is given by the boundary of anti-de Sitter space. This observation is the starting point for AdS/CFT correspondence, which states that the boundary of anti-de Sitter space can be regarded as the "spacetime" for a quantum field theory. The claim is that this quantum field theory is equivalent to the gravitational theory on the bulk anti-de Sitter space in the sense that there is a "dictionary" for translating entities and calculations in one theory into their counterparts in the other theory. For example, a single particle in the gravitational theory might correspond to some collection of particles in the boundary theory. In addition, the predictions in the two theories are quantitatively identical so that if two particles have a 40 percent chance of colliding in the gravitational theory, then the corresponding collections in the boundary theory would also have a 40 percent chance of colliding.

6D (2,0) Superconformal Field Theory

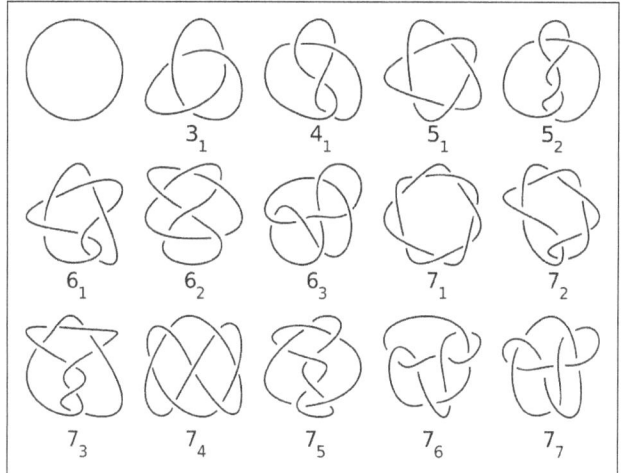

The six-dimensional (2,0)-theory has been used to understand results from the mathematical theory of knots.

One particular realization of the AdS/CFT correspondence states that M-theory on the product space $AdS_7 \times S^4$ is equivalent to the so-called (2,0)-theory on the six-dimensional boundary. Here "(2,0)" refers to the particular type of supersymmetry that appears in the theory. In this example, the spacetime of the gravitational theory is effectively seven-dimensional (hence the notation AdS_7), and there are four additional "compact" dimensions (encoded by the S^4 factor). In the real world, spacetime is four-dimensional, at least macroscopically, so this version of the correspondence does not provide a realistic model of gravity. Likewise, the dual theory is not a viable model of any real-world system since it describes a world with six spacetime dimensions.

Nevertheless, the (2,0)-theory has proven to be important for studying the general properties of quantum field theories. Indeed, this theory subsumes many mathematically interesting effective quantum field theories and points to new dualities relating these theories. For example, Luis Alday, Davide Gaiotto, and Yuji Tachikawa showed that by compactifying this theory on a surface, one obtains a four-dimensional quantum field theory, and there is a duality known as the AGT correspondence which relates the physics of this theory to certain physical concepts associated with the surface itself. More recently, theorists have extended these ideas to study the theories obtained by compactifying down to three dimensions.

In addition to its applications in quantum field theory, the (2,0)-theory has spawned important results in pure mathematics. For example, the existence of the (2,0)-theory was used by Witten to give a "physical" explanation for a conjectural relationship in mathematics called the geometric Langlands correspondence. In subsequent work, Witten showed that the (2,0)-theory could be used to understand a concept in mathematics called Khovanov homology. Developed by Mikhail Khovanov around 2000, Khovanov homology provides a tool in knot theory, the branch of mathematics that studies and classifies the different shapes of knots. Another application of the (2,0)-theory in mathematics is the work of Davide Gaiotto, Greg Moore, and Andrew Neitzke, which used physical ideas to derive new results in hyperkähler geometry.

ABJM Superconformal Field Theory

Another realization of the AdS/CFT correspondence states that M-theory on $AdS_4 \times S^7$ is equivalent to a quantum field theory called the ABJM theory in three dimensions. In this version of the correspondence, seven of the dimensions of M-theory are curled up, leaving four non-compact dimensions. Since the spacetime of our universe is four-dimensional, this version of the correspondence provides a somewhat more realistic description of gravity.

The ABJM theory appearing in this version of the correspondence is also interesting for a variety of reasons. Introduced by Aharony, Bergman, Jafferis, and Maldacena, it is closely related to another quantum field theory called Chern–Simons theory. The latter theory was popularized by Witten in the late 1980s because of its applications to knot theory. In addition, the ABJM theory serves as a semi-realistic simplified model for solving problems that arise in condensed matter physics.

Phenomenology

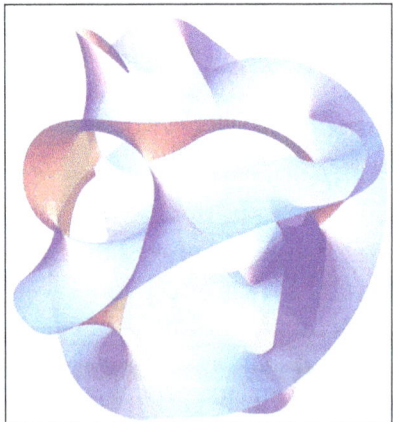

A cross section of a Calabi–Yau manifold.

In addition to being an idea of considerable theoretical interest, M-theory provides a framework for constructing models of real world physics that combine general relativity with the standard model of particle physics. Phenomenology is the branch of theoretical physics in which physicists construct realistic models of nature from more abstract theoretical ideas. String phenomenology is the part of string theory that attempts to construct realistic models of particle physics based on string and M-theory.

Typically, such models are based on the idea of compactification.[1] Starting with the ten- or eleven-dimensional spacetime of string or M-theory, physicists postulate a shape for the extra dimensions. By choosing this shape appropriately, they can construct models roughly similar to the standard model of particle physics, together with additional undiscovered particles, usually supersymmetric partners to analogues of known particles. One popular way of deriving realistic physics from string theory is to start with the heterotic theory in ten dimensions and assume that the six extra dimensions of spacetime are shaped like a six-dimensional Calabi–Yau manifold. This is a special kind of geometric object named after mathematicians Eugenio Calabi and Shing-Tung Yau. Calabi–Yau manifolds offer many ways of extracting realistic physics from string theory. Other similar methods can be used to construct models with physics resembling to some extent that of our four-dimensional world based on M-theory.

Partly because of theoretical and mathematical difficulties and partly because of the extremely high energies (beyond what is technologically possible for the foreseeable future) needed to test these theories experimentally, there is so far no experimental evidence that would unambiguously point to any of these models being a correct fundamental description of nature. This has led some in the community to criticize these approaches to unification and question the value of continued research on these problems.

Compacti ication on G_2 Manifolds

In one approach to M-theory phenomenology, theorists assume that the seven extra dimensions of M-theory are shaped like a G_2 manifold. This is a special kind of seven-dimensional shape constructed by mathematician Dominic Joyce of the University of Oxford. These G_2 manifolds are still poorly understood mathematically, and this fact has made it difficult for physicists to fully develop this approach to phenomenology.

For example, physicists and mathematicians often assume that space has a mathematical property called smoothness, but this property cannot be assumed in the case of a G_2 manifold if one wishes to recover the physics of our four-dimensional world. Another problem is that G_2 manifolds are not complex manifolds, so theorists are unable to use tools from the branch of mathematics known as complex analysis. Finally, there are many open questions about the existence, uniqueness, and other mathematical properties of G_2 manifolds, and mathematicians lack a systematic way of searching for these manifolds.

Heterotic M-theory

Because of the difficulties with G_2 manifolds, most attempts to construct realistic theories of physics based on M-theory have taken a more indirect approach to compactifying eleven-dimensional spacetime. One approach, pioneered by Witten, Hořava, Burt Ovrut, and others, is known as heterotic M-theory. In this approach, one imagines that one of the eleven dimensions of M-theory is shaped like a circle. If this circle is very small, then the spacetime becomes effectively ten-dimensional. One then assumes that six of the ten dimensions form a Calabi–Yau manifold. If this Calabi–Yau manifold is also taken to be small, one is left with a theory in four-dimensions.

Heterotic M-theory has been used to construct models of brane cosmology in which the observable universe is thought to exist on a brane in a higher dimensional ambient space. It has also spawned alternative theories of the early universe that do not rely on the theory of cosmic inflation.

LATTICE FIELD THEORY

Lattice field theory is an area of theoretical physics, specifically quantum field theory, which deals with field theories defined on a spatial or space-time lattice.

The theoretical description of the fundamental constituents of matter and the interactions between them is based on quantum field theory. The basic ingredients of field theory are fields. They are functions φ which associate to each point x of space-time a quantity φ(x) .

In the case of classical field theories, φ(x) usually is an element of a finite dimensional real or complex manifold, which in many cases is a linear space. Prominent examples are:

- The real scalar field $\phi(x) \in R$,

- The complex scalar field $\phi(x) \in C$,

- The n-vector field $\phi(x) \in Rn$,

- The photon field $A\mu(x) \in R$, where $\mu = 0,1,2,3$ is a Lorentz index,

- The Yang-Mills field $A_\mu(x) = \sum_b A_\mu^b(x)T_b$ whose components are elements of the Lie algebra of a compact Lie group with generators Tb.

In contrast, in the operator formulation of quantum field theory the fields are operators acting in a Hilbert space. (More precisely, quantum fields φ(x)are operator valued distributions, which means that integrals $\int f(x)\phi(x)dx$ with suitable test functions $f(x)$ are operators.).

The physical content of a field theory depends essentially on the Lagrangian $L(\phi(x), \partial n\phi(x))$, which is a function of φ(x) and its derivatives. The Lagrangian determines the field equations, which comprise the interactions. If the strength of an interaction is given by a small parameter g, it is possible to calculate physical quantities approximately to a satisfactory accuracy by means of perturbation theory, which amounts to a power series expansion in g. This is, for example, the case in quantum electrodynamics (QED), where the interaction is proportional to the fine structure constant $\alpha \approx 1/137$, and many interesting observables can be obtained as power series in α. There are, however, important cases, where it turned out that perturbation theory is inadequate for the calculation of physical quantities. The most prominent example is the low-energy regime of Quantum Chromodynamics (QCD), the theory of the strong interactions of elementary particles.

Not only Quantum Chromodynamics, but also other components of the Standard Model of elementary particle physics and moreover theories of physics beyond the Standard Model supply us with non-perturbative problems. An important step to answer such questions has been made by K. Wilson in 1974. He introduced a formulation of Quantum Chromodynamics on a space-time lattice, which allows the application of various non-perturbative techniques. It leads to mathematically well-defined problems, which are (at least in principle) solvable. It should also be pointed out that the introduction of a space-time lattice can be taken as a starting point for a mathematically clean approach to quantum field theory, so-called constructive quantum field theory.

In modern quantum field theory, the introduction of a space-time lattice is part of an approach different from the operator formalism. This is lattice field theory. Its main ingredients are:

- Functional integrals,

- Euclidean field theory,

- The space-time discretization of fields.

Lattice field theory has turned out to be very successful for the non-perturbative calculation of physical quantities. The main concepts are here illustrated with a scalar field theory.

Quantum Field Theory with Functional Integrals

The functional integral formulation of quantum field theory is a generalization of the quantum mechanical path integral. In quantum mechanics of a point particle in one space dimension, the transition amplitude is given by:

$$\langle x' | e^{-iHT} | x \rangle,$$

where, $|x\rangle$ is an (improper) eigenstate of the position operator and H is the Hamilton operator.

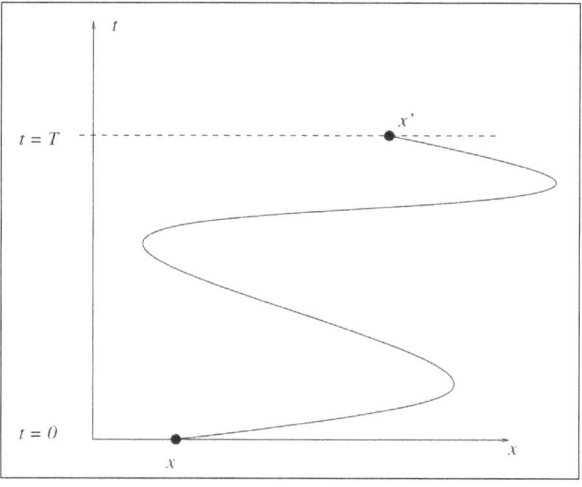

Path of a particle.

The transition amplitude can be written as a path integral:

$$\langle x' | e^{-iHT} | x \rangle = \int Dx \, e^{iS},$$

where the integration is over all possible paths $x(t)$ from x to x' during the time interval T , and
$$S = \int_0^T dt L(x, \dot{x})$$
s the classical action for such a path.

Formally the path integral measure is written as:

$$Dx \equiv \prod_t dx(t)$$

up to a normalization factor. For a particle in 3 dimensional space this is generalized to paths $x_{i(t)}$, where $i=1,2,3$, and

$$Dx \equiv \prod_t \prod_i dx_i(t).$$

Perhaps this is the most intuitive picture of the quantum mechanical transition amplitude. It can be written as an integral over contributions from all possible paths from the starting point to the final point. Each path is weighted by the classical action evaluated along this path.

The representation of quantum mechanics in terms of path integrals can be translated to field theory. Consider a scalar field $\varphi(x)$, where $x = (\vec{x}, t)$ labels space-time coordinates, and the time evolution of $\varphi(\vec{x}, t)$ is given by:

$$\varphi(\vec{x}, t) = e^{iHt} \varphi(\vec{x}, t = 0) e^{-iHt}.$$

The objects of interest in field theory are vacuum expectation values of (time ordered) products of field operators, the Greens functions:

$$\langle 0 | \varphi(x_1)\varphi(x_2)\ldots\varphi(x_n) | 0 \rangle, \quad t_1 > t_2 > \cdots > t_n.$$

Prominent examples are propagators $\langle 0 | \varphi(x)\varphi(y) | 0 \rangle$. The Greens functions essentially contain all physical information. In particular, S-matrix elements are related to Greens functions, e.g. the 2-particle scattering elements can be obtained from $\langle 0 | \varphi(x1)\ldots\varphi(x4) | 0 \rangle$.

Instead of discussing the functional integral representation for quantum field theory from the beginning, we shall restrict ourselves to translating the quantum mechanical concepts to field theory by means of analogy. To this end the basic variables $x_i(t)$ are translated into fields $\phi(\vec{x}, t)$. The rules for the translation are:

$$x_i(t) \leftrightarrow \phi(\vec{x}, t)$$
$$\quad\quad {}_{i \leftrightarrow \vec{x}}$$
$$\prod_{t,i} dx_{i(t)} \leftrightarrow \prod_{t,\vec{x}} d\phi(\vec{x}, t) \equiv D\phi$$
$$S = \int dt\, L \leftrightarrow S = \int dt\, d^3x L,$$

where S is the classical action.

For scalar field theory one might consider the following Lagrangian density:

$$L = \frac{1}{2}((\dot{\phi}(x))^2 - (\nabla\phi(x))^2) - \frac{m_0^2}{2}\phi(x)^2 - \frac{g0}{4!}\phi(x)^4$$
$$= \frac{1}{2}(\partial_\mu\phi)(\partial^\mu\phi) - \frac{m_0^2}{2}\phi(x)^2 - \frac{g0}{4!}\phi(x)^4.$$

The mass $m0$ and coupling constant $g0$ bear a subscript 0, since they are bare, unrenormalized parameters. This theory plays a role in the context of Higgs-Yukawa models, where $\phi(x)$ is the Higgs field.

In analogy to the quantum mechanical path integral, a representation of the Greens functions in terms of what one calls functional integrals is written down as:

$$\langle 0\,|\,\varphi(x1)\varphi(x2)...\varphi(xn)\,|\,0\rangle = \frac{1}{Z}\int D\phi\,\phi(x_1)\phi(x_2)...\phi(x_n)e^{iS}$$

With $Z = \int D\phi e^{iS}$ These expressions involve integrals over all classical field configurations.

As mentioned before, any derivation of functional integrals is not attempted here, but just a motivation of their form by analogy. Furthermore, in the case of quantum mechanics the transition amplitude has been considered, whereas now the formula for Greens functions has been written, which is a bit different.

The formulae for functional integrals give rise to some questions. First of all, how does the projection onto the ground state $|0\rangle$ arise? Secondly, these integrals contain oscillating integrands, due to the imaginary exponents; what about their convergence? Moreover, is there a way to evaluate them numerically?

Euclidean Field Theory

Return to quantum mechanics for a moment. Here one can also introduce Greens functions, e.g.

$$G(t_1,t_2) = \langle 0\,|\,X(t_1)X(t_2)\,|\,0\rangle,\ t_1 > t_2,$$

where $X(t)$ is the position operator in the Heisenberg picture. In the following it will be demonstrated that these Greens functions are related to quantum mechanical amplitudes at imaginary times by analytic continuation. Consider the matrix element:

$$\langle x',t'\,|\,X(t_1)X(t_2)\,|\,x,t\rangle = \langle x'\,|\,e^{-iH}(t'-t_1)Xe^{-iH}(t_1-t_2)Xe^{-iH(t_2-t)}\,|\,0\rangle$$

For $t' > t_1 > t_2 > t$. Now choose all times to be purely imaginary

again ordered, $\tau' > \tau_1 > \tau_2 > \tau$. This yields the expression.

Inserting a complete set of energy eigenstates, the expansion of the time evolution operator in imaginary times is $e^{-H\tau} = \sum_{n=0}^{\infty} e^{-E_n\tau}\,|\,n\rangle\langle n\,|=|\,0\rangle\langle 0\,|+e^{-E1\tau}\,|\,1\rangle\langle 1\,|+...,$

where the ground state energy has been normalized to $E_0 = 0$. For large τ it reduces to the projector onto the ground state. Consequently, in the limit $\tau' \to \infty$ and $\tau \to -\infty$ our matrix element becomes,

$$\langle x'\,|\,0\rangle\langle 0\,\|\,Xe^{-H(\tau_1-\tau_2)}X\,|\,0\rangle\langle 0\,|\,x\rangle,$$

and similarly,

$$\langle x'\,|\,e^{-H(\tau'-\tau)}\,|\,x\rangle \to \langle x'\,|\,0\rangle\langle 0\,|\,x\rangle.$$

Therefore, the Greens function at imaginary times,

$$G_E(\tau_1,\tau_2) = \langle 0\,|\,Xe^{-H(\tau_1-\tau_2)}X\,|\,0\rangle,$$

can be expressed as:

$$G_E(\tau_1,\tau_2) = \lim_{\tau'\to\infty,\tau\to-\infty} \frac{\langle x'|e^{-H(\tau'-\tau)}Xe^{-H(\tau_1-\tau_2)}Xe^{-H(\tau_2-\tau)}|x\rangle}{\langle x'|e^{-H(\tau'-\tau)}|x\rangle}$$

Now the denominator as well as the numerator can be represented by path integrals. The difference to the case of real times is that for imaginary times we have to use:

$$\langle x|e^{-H\Delta\tau}|y\rangle \approx \sqrt{\frac{m}{2\pi\Delta\tau}}exp-\Delta\tau\left\{\frac{m}{2}\right\}\left(\frac{x-y}{\Delta\tau}\right)^2 +V(x)\right\}.$$

This leads to the path integral representation:

$$G_E(\tau_1,\tau_2) = \frac{1}{Z}\int Dx\, x(\tau_1)x(\tau_2)e^{-SE},$$

where, $Z = \int Dx\, e^{-S_E}$ and,

The Greens function at real times, which we were interested in originally, can be obtained from GE by means of analytical continuation, $G(t1,t2)=GE(it1,it2)$.

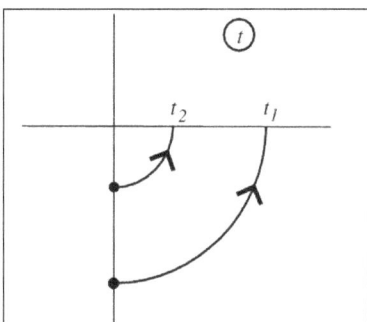

Wick rotation from imaginary to real time coordinates.

The analytic continuation has to be done in such a way that all time arguments are rotated simultaneously counter-clockwise in the complex t-plane. This is the so-called Wick rotation, illustrated in figure.

Now we turn to field theory again. The Green's functions:

$$G(x_1,\ldots,x_n) = \langle 0|T\varphi(x_1)\ldots\varphi(x_n)|0\rangle,$$

continued to imaginary times, $t=-i\tau$, are the so-called Schwinger functions:

$$G_E((\vec{x}_1,\tau_1),\ldots,(\vec{x}_n,\tau_n)) = G((\vec{x}_1,-i\tau_1),\ldots,(\vec{x}_n,-i\tau_n)).$$

In analogy to the quantum mechanical case their functional integral representation reads:

$$G_E(x_1,\ldots,x_n) = \frac{1}{Z}\int D\phi\,\phi(x_1)\ldots\phi(x_n)e^{-S_E}$$

with $Z = \int D\phi \, e^{-S_E}$

and

$$S_E = \int d^3x \, d\tau \left\{ \frac{1}{2} \left(\frac{d\phi}{d\tau} \right)^2 + \frac{1}{2} (\nabla \phi)^2 + \frac{m_0^2}{2} \phi^2 + \frac{g_0}{4_!} \phi^4 \right\}$$

$$= \int d^4x \left\{ \frac{1}{2} (\partial \mu \phi)^2 + \frac{m_0^2}{2} \phi^2 + \frac{g_0}{4_!} \phi^4 \right\}.$$

As can also be seen from the kinetic part contained in S_E, the metric of Minkowski space:

$$-ds^2 = -dt^2 + dx_1^2 + dx_2^2 + dx_3^2$$

has changed into:

$$d\tau^2 + dx_1^2 + dx_2^2 + dx_3^2,$$

which is the metric of a Euclidean space. Therefore one speaks of Euclidean Greens functions G_E and of Euclidean functional integrals. They are taken as starting point for non-perturbative investigations of field theories and for constructive studies.

Whether it is possible to continue a specific field theory analytically from real to imaginary times and vice versa, depends on certain conditions to be satisfied. For a large class of field theories these conditions have been analyzed and formulated by Osterwalder and Schrader,. In particular, a Euclidean field theory must satisfy the so-called reflection positivity in order to correspond to a proper field theory in Minkowski space.

As S_E is real, the integrals of interest are now real and no unpleasant oscillations occur. Moreover, since S_E is bounded from below, the factor $\exp(-S_E)$ in the integrand is bounded. Strongly fluctuating fields have a large Euclidean action S_E and are thus suppressed by the factor $\exp(-S_E)$. (Strictly speaking, this statement does not make sense in field theory unless renormalization is taken into account.) This makes Euclidean functional integrals so attractive compared to their Minkowskian counterparts.

One might think that in the Euclidean domain everything is unphysical and there is no possibility to get physical results directly from the Euclidean Greens functions. But this is not the case. For example, the spectrum of the theory can be obtained in the following way. Consider a vacuum expectation value of the form:

$$\langle 0 | A_1 e^{-H\tau} A_2 | 0 \rangle,$$

where the Ai's are formed out of the field φ, e.g. $A = \varphi(\vec{x}, 0)$ or $A = \int d^3x \, \varphi(\vec{x}, 0)$. Now, with the familiar insertion of a complete set of energy eigenstates, one has:

$$\langle 0 | A_1 e^{-H\tau} A_2 | 0 \rangle = \sum_n \langle 0 | A_1 | n \rangle e^{-E_n^\tau} \langle n | A2 | 0 \rangle.$$

In case of a continuous spectrum the sum is to be read as an integral. On the other hand, representing the expectation value as a functional integral leads to:

$$\frac{1}{Z}\int D\phi e^{-S_E} A_1(\tau)A_2(0) = \sum_n \langle 0|A_1|n\rangle\langle n|A2|0\rangle e^{-E_n^\tau}.$$

For large τ the lowest energy eigenstates will dominate the sum and one can thus obtain the low-lying spectrum from the asymptotic behaviour of this expectation value. By choosing A1, A2 suitably, e.g. for:

$$A \equiv A_1 = A_2 = \int \int d^3x\, \varphi(\vec{x},0),$$

such that $\langle 0|A|1\rangle \neq 0$ for a one-particle state $|1\rangle$ with zero momentum $\vec{p}=0$ and mass m1 , one gets,

$$\frac{1}{Z}\int D\phi e^{-S_E} A(\tau)A(0) = |\langle 0|A|1\rangle|^2 \, e - m_1^\tau + ...,$$

which means that one can extract the mass of the particle.

From now on we shall remain in Euclidean space and suppress the subscript E , so that $S \equiv SE$ means the Euclidean action.

Lattice Discretization

One central question still remains: does the infinite dimensional integration over all classical field configurations, i.e.

$$D\phi = \prod_x d\phi(x),$$

make sense at all? How is it defined?

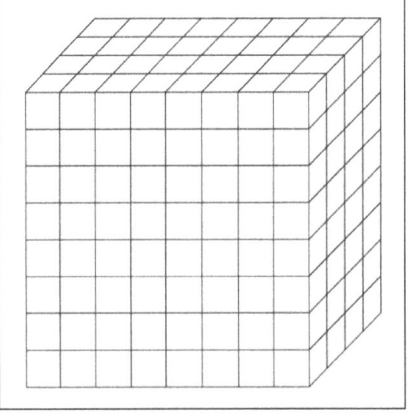

3-dimensional lattice.

In quantum mechanics the path integral representation can be derived as a limit of a discretization in time. As in field theory the fields depend on the four Euclidean coordinates instead of a single time coordinate, we may now introduce a discretized space-time in form of a lattice, for example a hypercubic lattice, specified by:

$$x_\mu = a n_\mu, \quad n_\mu \in Z,$$

The quantity a is called the lattice spacing for obvious reasons. It should be noted that the lattice spacing, being a dimensionful quantity, is not a parameter of the discretized theory, which could e.g. be inserted in a computer program for an evaluation of the path integral. The size of the lattice spacing in physical units is a derived quantity determined by the dynamics. This will be explained in "Continuum limit".

The scalar field,

$$\phi(x), \quad x \in \text{lattice},$$

is now defined on the lattice points only. Partial derivatives are replaced by finite differences,

$$\partial_\mu \phi \to \Delta_\mu \phi(x) \equiv \frac{1}{a}\left(\phi\left(x + a\hat{\mu}\right) - \phi(x)\right),$$

and space-time integrals by sums:

$$\int d^4 x \to \sum_x a^4.$$

The action of discretized φ⁴-theory, Eq.(1), can be written as:

$$S = \sum_x a^4 \left\{ \sum_{\mu=1}^{4} \left(\Delta_\mu \phi(x)\right)^2 + \frac{m_0^2}{2} \phi(x)^2 + \frac{g_0}{4!} \phi(x)^4 \right\}.$$

In the functional integrals the measure Dφ, Eq.(2), involves the lattice points x only. So a discrete set of variables has to be integrated. If the lattice is taken to be finite, one just has finite dimensional integrals.

Discretization of space-time using lattices has one very important consequence. Due to a non-zero lattice spacing, a cutoff in momentum space arises. The cutoff can be observed by having a look at the Fourier transformed field:

$$\tilde{\phi}(p) = \sum_x a^4 e^{-ipx} \phi(x).$$

The Fourier transformed functions are periodic in momentum-space, so that one can identify:

$$p_\mu \cong p_\mu + \frac{2\pi}{a}$$

and restrict the momenta to the so-called first Brillouin zone:

$$-\frac{\pi}{a} < p_\mu \le \frac{\pi}{a}.$$

The inverse Fourier transformation, for example, is given by:

$$\phi(x) = \int_{-\pi/a}^{\pi/a} \frac{d^4 p}{(2\pi)^4} e^{-ipx} \tilde{\phi}(p).$$

One recognises an ultraviolet cutoff:

$$\left| p_\mu \right| \leq \frac{\pi}{a}.$$

Therefore field theories on a lattice are regularized in a natural way.

In order to begin in a well-defined way one would start with a finite lattice. Let us assume a hypercubic lattice with length $L_1 = L_2 = L_3 = L$ in every spatial direction and length $L_4 = T$ in Euclidean time,

$$x_\mu = an_\mu, \quad n_\mu = 0, 1, 2, \ldots, L_\mu - 1,$$

with finite volume $V = L_3 T$. In a finite volume one has to specify boundary conditions. A popular choice are periodic boundary conditions:

$$\phi(x) = \phi\left(x + aL \; \hat{\mu}\right),$$

where $\hat{\mu}$ is the unit vector in the μ-direction. They imply that the momenta are also discretized,

$$p_\mu = \frac{2\pi}{a} \frac{l_\mu}{L_\mu} \quad \text{with } l_\mu = 0, 1, 2, \ldots, L_\mu - 1,$$

and therefore momentum-space integration is replaced by finite sums:

$$\int \frac{d^4 p}{(2\pi)^4} \quad \to \quad \frac{1}{a^4 L^3 T} \sum_{l_\mu}.$$

Now, all functional integrals have turned into regularized and finite expressions.

Of course, one would like to recover physics in a continuous and infinite space-time eventually. The task is therefore to take the infinite volume limit,

$$L, T \to \infty,$$

which is the easier part in general, and to take the continuum limit,

$$a \to 0.$$

Constructing the continuum limit of a lattice field theory is usually highly nontrivial and most effort is often spent here.

The formulation of Euclidean quantum field theory on a lattice bears a useful analogy to statistical mechanics. Functional integrals have the form of partition functions and we can set up the following correspondence:

Euclidean field theory	Statistical Mechanics
generating functional $\int D\phi\, e^{-S}$	partition function $\sum e^{-\beta H}$
action S	Hamilton function βH
mass m $G \sim e^{-mt}$	inverse correlation length $1/\xi$ $G \sim e^{-x\xi}$

This formal analogy allows to use well established methods of statistical mechanics in field theory and vice versa. Even the terminology of both fields is often identical. To mention some examples, in field theory one employs high-temperature expansions and mean field approximations, and in statistical mechanics one applies the renormalization group.

Hamiltonian Lattice Field Theory

An alternative to Euclidean lattice field theory, as described before, is Hamiltonian lattice field theory, introduced by Kogut and Susskind. In this formulation only three-dimensional space is discretized on a lattice, whereas time remains continuous. Furthermore, time is kept real and is not continued to the Euclidean domain. Hamiltonian lattice field theory allows the application of some analytical methods like strong coupling expansions and perturbation theory. Since it is not suitable for the application of the numerical Monte Carlo method, it doesn't enjoy any more as much attention as in its beginnings, and is not covered in more detail here.

Lattice Gauge Theory

Theories of gauge fields can also be formulated on a space-time lattice. We shall just indicate the basic elements of lattice gauge theory for gauge group SU(N).

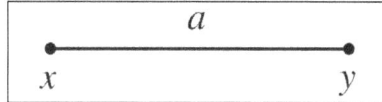

Link $b=\langle x,y \rangle$ between lattice points x and y.

The paths connecting nearest neighbour points on the lattice are called links.

With each link:

$$b = \left\langle x + a\widehat{\mu}, x \right\rangle$$

in lattice direction $\widehat{\mu}$ a link variable:

$$U(b) \equiv U(x + a\widehat{\mu}\,x) \equiv U_{x\mu} \in SU(N)$$

is associated. These group valued variables represent the gauge field. The discretized Lie algebra valued gauge field $A_\mu^b(x)$ can be introduced by:

$$U_{x\mu} \equiv \exp\{ig_0 a A_\mu^b(x)T_b\},$$

where the Ta's are generators of the gauge group and $g0$ is the bare coupling constant.

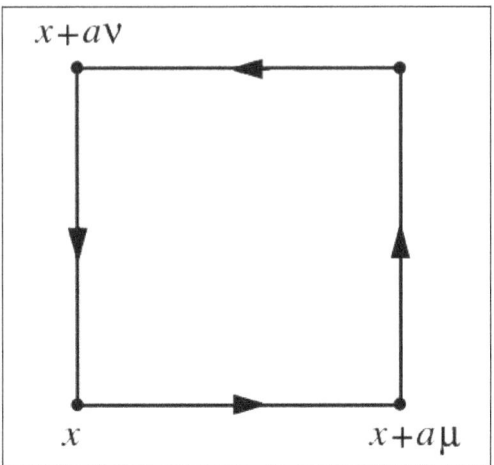

Plaquette p in lattice directions μ and v.

The smallest closed paths on the lattice are the plaquettes, as shown in figure.

The plaquette variables:

$$U(p) = U_{x\mu v} = U_{(x+a\hat{v})(-v)} U_{(x+a\hat{\mu}+a\hat{v})(-\mu)} U_{(x+a\hat{\mu})v} U_{x\mu}$$

enter the Wilson action:

$$S_W = -\sum_p \frac{2}{g_0^2} Re(Tr(U(p))).$$

In a naive continuum limit, where a goes to zero, one has:

$$S_W = \frac{g_0^2}{4} \sum_x a^4 F_{\mu v}^b F_{\mu v}^b + O(a^5),$$

which reproduces the Yang-Mills action.

The integral over all gauge field configurations on the lattice amounts to an integral over all link variables U(b). So, for the expectation value of any observable A one writes $\langle A \rangle = \frac{1}{Z} \int \prod_b dU(b) A e^{-SW}$,

where the integration dU(b) for a given link b is to be understood as the invariant integration over the group manifold, normalized to:

$$\int dU = 1.$$

Fermions on the Lattice

Grassmann Variables

Classical bosonic fields are just ordinary functions and satisfy:

$$[\phi(x),\phi(y)]=0,$$

which can be considered as the limit $\hbar \to 0$ of the quantum commutation relations.

Fermi statistics implies that fermionic quantum fields have the well-known equal-time anticommutation relations:

$$\{\psi(\vec{x},t),\psi(\vec{y},t)\}=0.$$

Motivated by this, one might introduce a classical limit in which classical fermionic fields satisfy:

$$\{\psi(x),\psi(y)\}=0$$

for all x,y . Classical fermionic fields are therefore anticommuting variables, which are also called Grassmann variables.

We would like to point out that the argument above is just a heuristic motivation. More rigorous approaches can be found in the literature.

In general, a complex Grassmann algebra is generated by elements ηi and $\bar{\eta} i$, which obey:

$$\{\eta_i,\eta_j\}=0$$
$$\{\eta_i,\eta_j\}=0$$
$$\{\eta_i,\eta_j\}=0.$$

n integration of Grassmann variables can be defined by

for arbitrary complex numbers a,b .

In fermionic field theories one has Grassmann fields, which associate Grassmann variables with every space-time point. For example, a Dirac field has anticommuting variables $\psi\alpha(x)$ and $\bar{\psi}\alpha(x)$, where α=1,2,3,4 is the Dirac index. The classical Dirac field obeys $\{\psi_\alpha(x),\psi_\beta(y)\}=0, \quad etc..$

In order to write down fermionic path integrals as integrals over fermionic and anti-fermionic field configurations, we write $D\psi D\bar{\psi}=\prod_x \prod_a d\psi_\alpha(x)d\bar{\psi}_\alpha(x).$

Then any fermionic Greens function is of the form:

$$\langle 0|A|0\rangle=\frac{1}{Z}\int D\psi D\bar{\psi}\, A e^{-S_F},$$

with an action SF for the fermions. For a free Dirac field the action is:

$$S_F=\int d^4x\,\bar{\psi}(x)(\gamma_\mu\partial^\mu+m)\psi(x).$$

In the context of the Standard Model, fermionic actions are always bilinear in the fermionic fields. With the help of the Grassmann integration rules above one can then show that the functional integrals are formally remarkably simple to calculate:

$$\int D\psi \, D\bar{\psi} e^{-\int d^4x \bar{\Psi}(x)Q\Psi(x)} = det \, Q.$$

This is the famous fermion determinant. The main problem remains, of course, namely to evaluate the determinant of the typically huge matrix Q .

In numerical simulations of lattice field theories with fermions the calculation of detQ turns out to be very tedious. Therefore one often uses the quenched approximation that treats Q as a constant. In recent years different unquenched investigations of Quantum Chromodynamics have been made and have given estimates for quenching errors.

Naive Fermions

So far no difficulties for the implementation of fermions on the lattice seem to arise: all one has to do is to discretise the field configurations in the well-known way and to calculate the Greens functions with some of the methods. There is a problem, however. The consider the propagator of a fermion with mass m as an example. The fermionic lattice action is then given by:

$$S_F = \frac{1}{2}\sum_x \sum_\mu \bar{\psi}(x)(\gamma_\mu \Delta_\mu + m)\psi(x) + h.c.$$

and the resulting propagator is:

$$\tilde{\Delta}(k) = \frac{-i\sum_\mu \gamma_\mu sink_\mu + m}{\sum_\mu sink_\mu^2 + m^2}.$$

The propagator has got a pole for small $k\mu$ representing the physical particle, but there are additional poles near $k\mu=\pm\pi$ due to the periodicity of the denominator. So *SF* really describes 16 instead of 1 particle. This problem - euphemistically called fermion doubling - is a crucial obstacle for all lattice representations of quark fields.

Wilson and Staggered Fermions

Fermion doubling was already known to Wilson in the early days of lattice Quantum Chromodynamics. He proposed a modified action for the fermions in order to damp out the doubled fields in the continuum limit. Therefore he added another term, the Wilson term, to the naive action.

$$S_F \rightarrow S_F^{(W)} = S_F - \frac{r}{2}\sum_x \bar{\psi}(x)\Box\psi(x)$$

$$= S_F - \frac{r}{2}\sum_{x,\mu}\bar{\psi}(x)\{\psi(x+\hat{\mu})+\psi(x-\hat{\mu})-2\psi(x)\},$$

where $0<r\leq1$. Calculating the propagator with this modified action, one finds that the unwanted doubled fermions acquire masses $\propto1/a$, so that they become infinitely massive in the continuum limit and disappear from the physical spectrum.

Wilson fermions have a serious disadvantage: even at vanishing fermion masses, chiral symmetry is broken explicitly by the Wilson term, and one has problems with calculations for which chiral symmetry is of central importance.

There are alternatives to Wilson's approach. One of them, due to Kogut and Susskind, are so-called staggered fermions. The idea is to distribute the components ψa of the Dirac field on different lattice points. It results in a reduction from 16 to 4 fermions. Moreover, for massless fermions a remnant of chiral symmetry in form of a chiral U(1)⊗U(1)-symmetry remains.

Even better in view of chiral symmetry and other aspects are formulations for fermions on the lattice, which obey the Ginsparg-Wilson relation. More details can be found in the Wikis on lattice gauge theories and on lattice chiral fermions.

Methods

In the previous sections the functional integrals for field theories on the lattice have been defined. But it is another problem to evaluate these high dimensional integrals. A calculation in closed form appears to be impossible in general. In this topic some of the methods used to evaluate the functional integrals approximately are considered.

Perturbation Theory

Although lattice field theory offers the possibility to study non-perturbative aspects, perturbation theory is nevertheless a highly valuable tool on the lattice, too. In particular, it can be used to match the results of non-perturbative calculations to perturbative calculations in regions where both methods are applicable.

Perturbation theory amounts to an expansion in powers of the coupling as in the continuum. The lattice provides an intrinsic UV cutoff π/a for all momenta. Apart from that one has to observe that the propagators and vertices are different from the continuum ones, owing to the form of the lattice action. In particular, gluon self interactions of all orders appear and not only as three and four gluon vertices.

Strong Coupling Expansion

The analogies between Euclidean field theory and statistical mechanics have already been pointed out. In statistical mechanics a well-established technique is the high-temperature expansion. For lattice gauge theory, this is an expansion in powers of $\beta \sim \dfrac{1}{g_0^2}$, which is a small quantity at large bare couplings g0 . Therefore it is the same as a strong coupling expansion. Basically the Boltzmann factor is expanded as:

$$\exp(\beta \frac{1}{N} Re(Tr(U(p)))) = 1 + \beta \frac{1}{N} Re(Tr(U(p))) + \dots .$$

The resulting expansion can be represented diagrammatically, similar to the Feynman diagrams of perturbation theory. The diagram elements, however, are plaquettes p on the lattice. Every power of β introduces one more plaquette.

In the case of scalar fields, the corresponding method is the hopping parameter expansion, which amounts to an expansion in a parameter κ, which is small for large masses m_0.

Strong coupling and hopping parameter expansions have a finite radius of convergence, in contrast to perturbation theory, which usually is divergent and at best asymptotic.

Other Analytic Methods

Other analytical methods are available for approximative evaluations of the functional integrals of lattice gauge theory. Some of them are:

- Mean field approximation.

- Renormalization group.

- $\frac{1}{N}$-expansion.

Monte Carlo Methods

On a finite lattice the calculation of expectation values requires the evaluation of finite dimensional integrals. This immediately suggests the application of numerical methods. The first thing one would naively propose is some simple numerical quadrature. In order to understand that this approach wouldn't be all that helpful, consider a typical lattice as it is considered in recent calculations. With 40 lattice points in every direction we have $4 \cdot 40^4$ link variables. For gauge group SU(3) this gives 81,920,000 real variables. That should be intractable for conventional quadratures even in the future. Therefore some statistical method is required. Producing lattice gauge configurations just randomly turns out to be extremely inefficient. The crucial idea to handle this problem is the concept of importance sampling: for a given lattice action S quadrature points xi are generated with a probability:

$$p(x_i) \sim \exp\{-S(x_i)\}.$$

This provides us with a large number of points in the important regions of the integral, improving the accuracy drastically.

In case of lattice gauge theory the quadrature points are configurations $U(i) = \{U_{x\mu}^{(i)}\}$. An expectation value:

$$\langle 0|A|0\rangle = \frac{1}{Z}\int DU\, A(U)e^{-S(U)}$$

is numerically approximated by the average:

$$\overline{A} \equiv \frac{1}{n}\sum_{i=1}^{n} A(U^{(i)}).$$

The Monte Carlo method consists in producing a sequence of configurations $U(1)\rightarrow U(2)\rightarrow U(3)\rightarrow\ldots$ with the appropriate probabilities in a statistical way. This is of course done on a computer. An update is a step where a single link variable $Ux\mu$ is changed, whereas a sweep implies that one goes once through the entire lattice, updating all link variables. A commonly used technique for obtaining updates is the Metropolis algorithm.

An important feature of this statistical way of evaluation is the existence of statistical errors. The result of such a calculation is usually presented in the form:

$$\langle A \rangle = \overline{A} \pm \sigma_{\overline{A}},$$

where the variance of $\overline{A}$ decreases with the number n of configurations as:

$$\sigma_{\overline{A}} \sim \frac{1}{n^{1/2}}.$$

Error Sources

The results obtained by means of the Monte Carlo method differ from the desired physical results by different sorts of errors. The most important error sources are

- Statistical errors: due to the finite number of configurations in the Monte Carlo calculation, $\sim 1/n1/2$,

- Lattice effects: due to finite lattice spacing a , often $\sim$a or a2 ,

- Volume effects: due to finite lattice volume, often $\sim$1/L , 1/L2 , or e$-$mL ,

- Large quark masses

    m$_q$

    mostly too big in Monte Carlo calculations,

- Quenched approximation

    detQ=1 , neglecting the fermion dynamics.

Continuum Limit

As one is only able to perform calculations at finite lattice spacing, it is an important issue to get the extrapolation process to the continuum limit under control. Since the lattice spacing is the regulator of the theory, it should be useful to apply renormalization group techniques to this problem. Knowing the functional dependence of the bare coupling $g$on the regulator, in other words solving the renormalization group equation, we should know how to vary the bare coupling of our theory in order to reach a continuum limit. Let us discuss this idea in more detail.

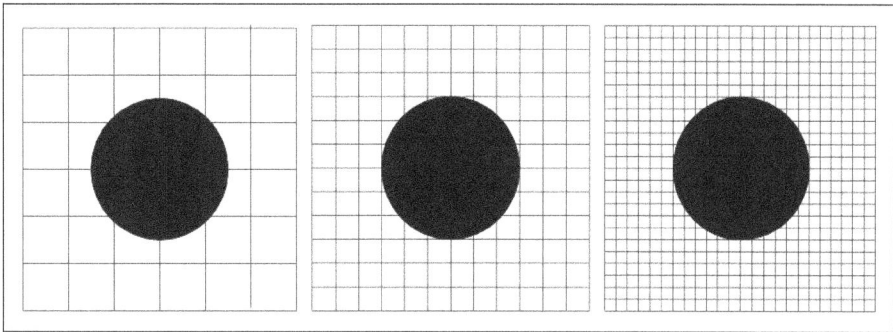

2-dimensional lattices with increasing correlation lengths ξ.

In the continuum limit the lattice spacing a is supposed to go to zero, while physical masses $m$should approach a finite limit. The lattice spacing, however, is not a dimensionless quantity, therefore we have to fix some mass scale m , e.g. some particle mass, and consider the limit $am\to0$. The inverse of that,

$$\frac{1}{am} \equiv \xi,$$

can be regarded as a correlation length. In the continuum limit ξ has to go to infinity, which is called a critical point of the theory. In figure this is illustrated on a two-dimensional lattice with different correlation lengths.

In pure gauge theory, there is a single, dimensionless bare coupling $g0$ and am is clearly a function of $g0$. In order to approach the continuum limit, we have to vary $g0$ such that $am\to0$. How this is done, is controlled by a renormalization group equation:

$$-a\frac{\partial g_0}{\partial_a} = \beta_{LAT}(g_0) = -\beta_0 g_0^3 - \beta_1 g_0^5 + ...,$$

where the first term of the expansion is:

$$\beta_0 = \frac{11}{3} N \frac{1}{16\pi^2}$$

In the perturbative regime of $g0$ this equation implies that for decreasing am the bare coupling $g0$ is also decreasing, getting even closer to zero. Hence the continuum limit is associated with the limit $g0\to0$(continuum limit).

The solution of the renormalization group equation up to second order in $g0$ is:

$$a = \Lambda_{LAT}^{-1} \exp\left(-\frac{1}{2\beta_0 g_0^2}\right)\left(\beta_0 g_0^2\right)^{-\frac{\beta_1}{2\beta_0^2}}\left\{1 + O\left(g_0^2\right)\right\},$$

where the lattice Λ-parameter ΛLAT appears. Solving for $g0$ yields:

$$g_0^2 = \frac{-1}{\beta_0 \log a^2 \Lambda_{LAT}^2} + ...,$$

which again reveals the vanishing of $g0$ in the continuum limit:

$$g_0^2 \to 0 \, \text{for} \, a \to 0.$$

We can also observe that:

$$am = C\exp\left(-\frac{1}{2\beta_0 g_0^2}\right)\cdot(...),$$

which shows the non-perturbative origin of the mass m.

These considerations, based on the perturbative β-function, motivate the following hypothesis: the continuum limit of a gauge theory on a lattice is to be taken at $g_0 \to 0$. Moreover, we expect that it involves massive interacting glueballs and static quark confinement.

The scenario for approaching the continuum limit then is as follows. Calculating masses in lattice units, i.e. numbers am, and decreasing g_0, we should reach a region where dimensionless quantities am follow a behaviour as given by above equation, which is called asymptotic scaling.

For mass ratios it can be shown that the exponential dependence on $1/g_0^2$ cancels out and it is thought that near the continuum limit:

$$\frac{m_1}{m_2} = \text{const.} \times (1 + O(a^p))$$

for some integer p. Such a behaviour, $m_1/m_2 \approx$ const., is called scaling. In numerical simulations scaling of various physical quantities has been established for lattice gauge theories, lattice QCD and other models, whereas confirmation of asymptotic scaling is much more demanding.

References

- Birnholtz, Ofek; Hadar, Shahar; Kol, Barak (2014). "Radiation reaction at the level of the action". International Journal of Modern Physics A. 29 (24): 1450132. Arxiv:1402.2610. Doi:10.1142/S0217751X14501322

- Quantum-field-theory, entries: plato.stanford.edu, Retrieved 19 July, 2019

- Birnholtz, Ofek; Hadar, Shahar; Kol, Barak (2014). "Radiation reaction at the level of the action". International Journal of Modern Physics A. 29 (24): 1450132. arXiv:1402.2610. doi:10.1142/S0217751X14501322

- Lattice-quantum-field-theory: scholarpedia.org, Retrieved 18 June, 2019

- Khovanov, Mikhail (2000). "A categorification of the Jones polynomial". Duke Mathematical Journal. 1011 (3): 359–426. arXiv:math/9908171. doi:10.1215/S0012-7094-00-10131-7

Laws of Physics

Laws of physics are derived from emperical observations. Laws such as Stefan-Boltzmann law, Pascal's law, Hooke's law, Charles's law and Boyle's law are among the basic laws of physics. The topics elaborated in this chapter will help in gaining a better perspective about these laws of physics.

STEFAN-BOLTZMANN LAW

Stefan-Boltzmann law states that the total radiant heat energy emitted from a surface is proportional to the fourth power of its absolute temperature. Formulated in 1879 by Austrian physicist Josef Stefan as a result of his experimental studies, the same law was derived in 1884 by Austrian physicist Ludwig Boltzmann from thermodynamic considerations: if E is the radiant heat energy emitted from a unit area in one second and T is the absolute temperature (in degrees Kelvin), then $E = \sigma T_4$, the Greek letter sigma (σ) representing the constant of proportionality, called the Stefan-Boltzmann constant. This constant has the value 5.670367×10^{-8} watt per metre^2 per K_4. The law applies only to blackbodies, theoretical surfaces that absorb all incident heat radiation.

PASCAL'S LAW

Pressure is defined as force per unit area. Can pressure be increased in a fluid by pushing directly on the fluid? Yes, but it is much easier if the fluid is enclosed. The heart, for example, increases blood pressure by pushing directly on the blood in an enclosed system (valves closed in a chamber). If you try to push on a fluid in an open system, such as a river, the fluid flows away. An enclosed fluid cannot flow away, and so pressure is more easily increased by an applied force. What happens to a pressure in an enclosed fluid? Since atoms in a fluid are free to move about, they transmit the pressure to all parts of the fluid and to the walls of the container. Remarkably, the pressure is transmitted undiminished. This phenomenon is called Pascal's principle, because it was first clearly stated by the French philosopher and scientist Blaise Pascal: A change in pressure applied to an enclosed fluid is transmitted undiminished to all portions of the fluid and to the walls of its container.

Pascal's principle, an experimentally verified fact, is what makes pressure so important in fluids. Since a change in pressure is transmitted undiminished in an enclosed fluid, we often know more about pressure than other physical quantities in fluids. Moreover, Pascal's principle implies that the total pressure in a fluid is the sum of the pressures from different sources. We shall find this fact—that pressures add—very useful.

Blaise Pascal had an interesting life in that he was home-schooled by his father who removed all of the mathematics textbooks from his house and forbade him to study mathematics until the age of 15. This, of course, raised the boy's curiosity, and by the age of 12, he started to teach himself geometry. Despite this early deprivation, Pascal went on to make major contributions in the mathematical fields of probability theory, number theory, and geometry. He is also well known for being the inventor of the first mechanical digital calculator, in addition to his contributions in the field of fluid statics.

Application of Pascal's Principle

One of the most important technological applications of Pascal's principle is found in a hydraulic system, which is an enclosed fluid system used to exert forces. The most common hydraulic systems are those that operate car brakes. Let us first consider the simple hydraulic system shown in figure.

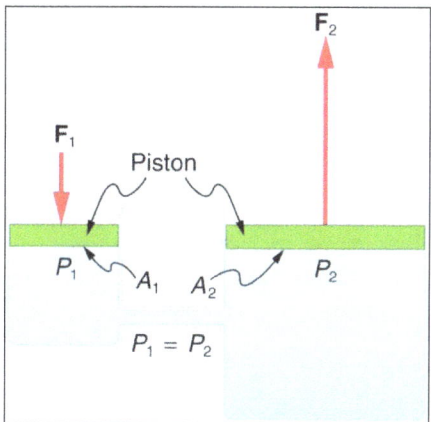

A typical hydraulic system with two fluid-filled cylinders, capped with pistons and connected by a tube called a hydraulic line.

A downward force F1 on the left piston creates a pressure that is transmitted undiminished to all parts of the enclosed fluid. This results in an upward force F2 on the right piston that is larger than F1 because the right piston has a larger area.

Relationship between Forces in a Hydraulic System

We can derive a relationship between the forces in the simple hydraulic system shown in Figure 1 by applying Pascal's principle. Note first that the two pistons in the system are at the same height, and so there will be no difference in pressure due to a difference in depth. Now the pressure due to F1 acting on area A1 is simply:

$$P_1 = \frac{F_1}{A_1}$$

as defined by:

$$P = \frac{F}{A}$$

According to Pascal's principle, this pressure is transmitted undiminished throughout the fluid and to all walls of the container. Thus, a pressure P2 is felt at the other piston that is equal to P1. That is $P_1 = P_2$. But since:

$$P_2 = \frac{F_2}{A_2}$$

we see that:

$$\frac{F_1}{A_1} = \frac{F_2}{A_2}$$

This equation relates the ratios of force to area in any hydraulic system, providing the pistons are at the same vertical height and that friction in the system is negligible. Hydraulic systems can increase or decrease the force applied to them. To make the force larger, the pressure is applied to a larger area. For example, if a 100-N force is applied to the left cylinder in Figure 1 and the right one has an area five times greater, then the force out is 500 N. Hydraulic systems are analogous to simple levers, but they have the advantage that pressure can be sent through tortuously curved lines to several places at once.

Example. Calculating force of slave cylinders: Pascal puts on the brakes.

Consider the automobile hydraulic system shown in figure:

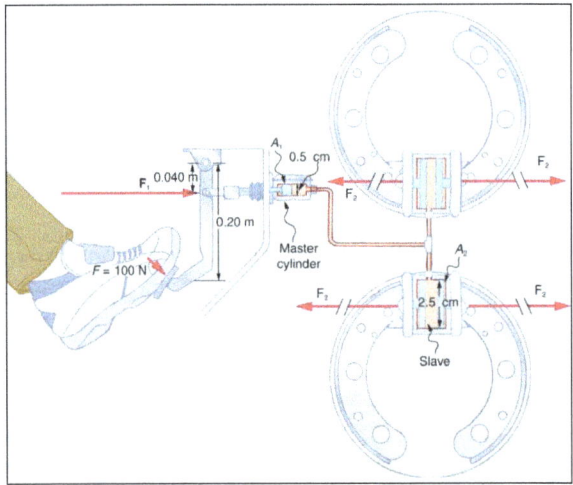

Hydraulic brakes use Pascal's principle. The driver exerts a force of 100 N on the brake pedal.

This force is increased by the simple lever and again by the hydraulic system. Each of the identical slave cylinders receives the same pressure and, therefore, creates the same force output F_2. The circular cross-sectional areas of the master and slave cylinders are represented by A_1 and A_2, respectively.

A force of 100 N is applied to the brake pedal, which acts on the cylinder—called the master—through a lever. A force of 500 N is exerted on the master cylinder. (The reader can verify that the force is 500 N using techniques of statics from Applications of Statics, including Problem-Solving Strategies.). Pressure created in the master cylinder is transmitted to four so-called slave cylinders.

The master cylinder has a diameter of 0.500 cm, and each slave cylinder has a diameter of 2.50 cm. Calculate the force F2 created at each of the slave cylinders.

Strategy

We are given the force F1 that is applied to the master cylinder. The cross-sectional areas A1 and A2 can be calculated from their given diameters. Then:

$$\frac{F_1}{A_1} = \frac{F_2}{A_2}$$

can be used to find the force F2. Manipulate this algebraically to get F2 on one side and substitute known values.

Solution

Pascal's principle applied to hydraulic systems is given by:

$$\frac{F_1}{A_1} = \frac{F_2}{A_2}$$

$$F_2 = \frac{A_2}{A_1} F_1 = \frac{\pi r_2^2}{\pi r_1^2} F_1 = \frac{(1.25\,cm)^2}{(0.250\,cm)^2} \times 500 \ \text{N} = 1.25 \times 10^4 \ \text{N}$$

This value is the force exerted by each of the four slave cylinders. Note that we can add as many slave cylinders as we wish. If each has a 2.50-cm diameter, each will exert 1.25×10^4 N.

A simple hydraulic system, such as a simple machine, can increase force but cannot do more work than done on it. Work is force times distance moved, and the slave cylinder moves through a smaller distance than the master cylinder. Furthermore, the more slaves added, the smaller the distance each moves. Many hydraulic systems—such as power brakes and those in bulldozers—have a motorized pump that actually does most of the work in the system. The movement of the legs of a spider is achieved partly by hydraulics. Using hydraulics, a jumping spider can create a force that makes it capable of jumping 25 times its length.

HOOKE'S LAW

Hooke's law is a law of physics that states that the force (F) needed to extend or compress a spring by some distance x scales linearly with respect to that distance. That is: $f_s = kx$, where k is a constant factor characteristic of the spring: its stiffness, and x is small compared to the total possible deformation of the spring. The law is named after 17th-century British physicist Robert Hooke. He first stated the law in 1676 as a Latin anagram. He published the solution of his anagram in 1678 as: *ut tensio, sic vis* ("as the extension, so the force" or "the extension is proportional to the force"). Hooke states in the 1678 work that he was aware of the law already in 1660.

Hooke's equation holds (to some extent) in many other situations where an elastic body is deformed, such as wind blowing on a tall building, and a musician plucking a string of a guitar. An elastic body or material for which this equation can be assumed is said to be linear-elastic or Hookean.

Hooke's law is only a first-order linear approximation to the real response of springs and other elastic bodies to applied forces. It must eventually fail once the forces exceed some limit, since no material can be compressed beyond a certain minimum size, or stretched beyond a maximum size, without some permanent deformation or change of state. Many materials will noticeably deviate from Hooke's law well before those elastic limits are reached.

On the other hand, Hooke's law is an accurate approximation for most solid bodies, as long as the forces and deformations are small enough. For this reason, Hooke's law is extensively used in all branches of science and engineering, and is the foundation of many disciplines such as seismology, molecular mechanics and acoustics. It is also the fundamental principle behind the spring scale, the manometer, and the balance wheel of the mechanical clock.

The modern theory of elasticity generalizes Hooke's law to say that the strain (deformation) of an elastic object or material is proportional to the stress applied to it. However, since general stresses and strains may have multiple independent components, the "proportionality factor" may no longer be just a single real number, but rather a linear map (a tensor) that can be represented by a matrix of real numbers.

In this general form, Hooke's law makes it possible to deduce the relation between strain and stress for complex objects in terms of intrinsic properties of the materials it is made of. For example, one can deduce that a homogeneous rod with uniform cross section will behave like a simple spring when stretched, with a stiffness k directly proportional to its cross-section area and inversely proportional to its length.

Formal Definition

For Linear Springs

Consider a simple helical spring that has one end attached to some fixed object, while the free end is being pulled by a force whose magnitude is F_s. Suppose that the spring has reached a state of equilibrium, where its length is not changing anymore. Let x be the amount by which the free end of the spring was displaced from its "relaxed" position (when it is not being stretched). Hooke's law states that:

$$F_s = kx$$

or, equivalently,

$$x = \frac{F_s}{k}$$

where k is a positive real number, characteristic of the spring. Moreover, the same formula holds when the spring is compressed, with F_s and x both negative in that case. According to this formula, the graph of the applied force F_s as a function of the displacement x will be a straight line passing through the origin, whose slope is k.

Hooke's law for a spring is often stated under the convention that F_s is the restoring force exerted by the spring on whatever is pulling its free end. In that case, the equation becomes:

$$F_s = -kx$$

since the direction of the restoring force is opposite to that of the displacement.

General "Scalar" Springs

Hooke's spring law usually applies to any elastic object, of arbitrary complexity, as long as both the deformation and the stress can be expressed by a single number that can be both positive and negative.

For example, when a block of rubber attached to two parallel plates is deformed by shearing, rather than stretching or compression, the shearing force F_s and the sideways displacement of the plates x obey Hooke's law (for small enough deformations).

Hooke's law also applies when a straight steel bar or concrete beam, supported at both ends, is bent by a weight F placed at some intermediate point. The displacement x in this case is the deviation of the beam, measured in the transversal direction, relative to its unloaded shape.

The law also applies when a stretched steel wire is twisted by pulling on a lever attached to one end. In this case the stress F_s can be taken as the force applied to the lever, and x as the distance traveled by it along its circular path. Or, equivalently, one can let F_s be the torque applied by the lever to the end of the wire, and x be the angle by which that end turns. In either case F_s is proportional to x (although the constant k is different in each case.).

Vector Formulation

In the case of a helical spring that is stretched or compressed along its axis, the applied (or restoring) force and the resulting elongation or compression have the same direction (which is the direction of said axis). Therefore, if F_s and x are defined as vectors, Hooke's equation still holds and says that the force vector is the elongation vector multiplied by a fixed scalar.

General Tensor Form

Some elastic bodies will deform in one direction when subjected to a force with a different direction. One example is a horizontal wood beam with non-square rectangular cross section that is bent by a transverse load that is neither vertical nor horizontal. In such cases, the *magnitude* of the displacement x will be proportional to the magnitude of the force F_s, as long as the direction of the latter remains the same (and its value is not too large); so the scalar version of Hooke's law $F_s = -kx$ will hold. However, the force and displacement *vectors* will not be scalar multiples of each other, since they have different directions. Moreover, the ratio k between their magnitudes will depend on the direction of the vector F_s.

Yet, in such cases there is often a fixed linear relation between the force and deformation vectors, as long as they are small enough. Namely, there is a function κ from vectors to vectors, such that $\mathbf{F} = \kappa(\mathbf{X})$, and $\kappa(\alpha\mathbf{X}_1 + \beta\mathbf{X}_2) = \alpha\kappa(\mathbf{X}_1) + \beta\kappa(\mathbf{X}_2)$ for any real numbers α, β and any displacement vectors $\mathbf{X}_1$, $\mathbf{X}_2$. Such a function is called a (second-order) tensor.

With respect to an arbitrary Cartesian coordinate system, the force and displacement vectors can be represented by 3 × 1 matrices of real numbers. Then the tensor **κ** connecting them can be represented by a 3 × 3 matrix **κ** of real coefficients, that, when multiplied by the displacement vector, gives the force vector:

$$F = \begin{bmatrix} F_1 \\ F_2 \\ F_3 \end{bmatrix} = \begin{bmatrix} \kappa_{11} & \kappa_{12} & \kappa_{13} \\ \kappa_{21} & \kappa_{22} & \kappa_{23} \\ \kappa_{31} & \kappa_{32} & \kappa_{33} \end{bmatrix} \begin{bmatrix} X_1 \\ X_2 \\ X_3 \end{bmatrix} = \kappa X$$

That is,

$$F_i = \kappa_{i1} X_1 + \kappa_{i2} X_2 + \kappa_{i3} X_3$$

for i = 1, 2, 3. Therefore, Hooke's law F = κX can be said to hold also when X and F are vectors with variable directions, except that the stiffness of the object is a tensor κ, rather than a single real number k.

Hooke's Law for Continuous Media

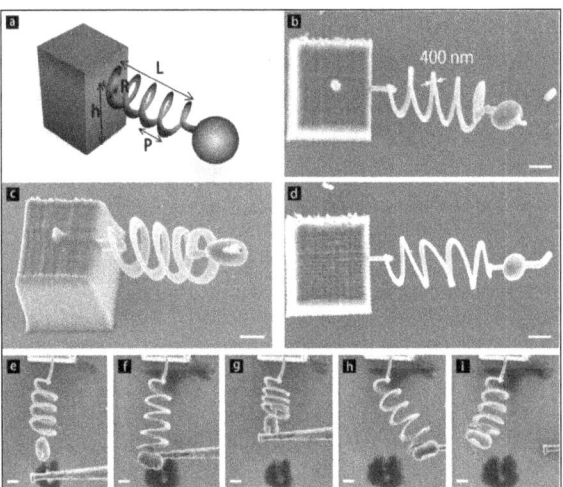

(a) Schematic of a polymer nanospring. The coil radius, R, pitch, P, length of the spring, L, and the number of turns, N, are 2.5 μm, 2.0 μm, 13 μm, and 4, respectively. Electron micrographs of the nanospring, before loading (b-e), stretched (f), compressed (g), bent (h), and recovered (i). All scale bars are 2 μm. The spring followed a linear response against applied force, demonstrating the validity of Hooke's law at the nanoscale.

The stresses and strains of the material inside a continuous elastic material (such as a block of rubber, the wall of a boiler, or a steel bar) are connected by a linear relationship that is mathematically similar to Hooke's spring law, and is often referred to by that name.

However, the strain state in a solid medium around some point cannot be described by a single vector. The same parcel of material, no matter how small, can be compressed, stretched, and sheared at the same time, along different directions. Likewise, the stresses in that parcel can be at once pushing, pulling, and shearing.

In order to capture this complexity, the relevant state of the medium around a point must be represented by two-second-order tensors, the strain tensor $\boldsymbol{\varepsilon}$ (in lieu of the displacement $\mathbf{X}$) and the stress tensor $\boldsymbol{\sigma}$ (replacing the restoring force $\mathbf{F}$). The analogue of Hooke's spring law for continuous media is then:

$$\sigma = -c\varepsilon,$$

where $\mathbf{c}$ is a fourth-order tensor (that is, a linear map between second-order tensors) usually called the stiffness tensor or elasticity tensor. One may also write it as:

$$\varepsilon = -s\sigma,$$

where the tensor $\mathbf{s}$, called the compliance tensor, represents the inverse of said linear map.

In a Cartesian coordinate system, the stress and strain tensors can be represented by 3×3 matrices:

$$\varepsilon = \begin{bmatrix} \varepsilon_{11} & \varepsilon_{12} & \varepsilon_{13} \\ \varepsilon_{21} & \varepsilon_{22} & \varepsilon_{23} \\ \varepsilon_{31} & \varepsilon_{32} & \varepsilon_{33} \end{bmatrix}; \qquad \sigma = \begin{bmatrix} \sigma_{11} & \sigma_{12} & \sigma_{13} \\ \sigma_{21} & \sigma_{22} & \sigma_{23} \\ \sigma_{31} & \sigma_{32} & \sigma_{33} \end{bmatrix}$$

Being a linear mapping between the nine numbers σ_{ij} and the nine numbers ε_{kl}, the stiffness tensor $\mathbf{c}$ is represented by a matrix of $3 \times 3 \times 3 \times 3 = 81$ real numbers c_{ijkl}. Hooke's law then says that:

$$\sigma_{ij} = -\sum_{k=1}^{3} \sum_{l=1}^{3} c_{ijkl} \varepsilon_{kl}$$

where $i,j = 1,2,3$.

All three tensors generally vary from point to point inside the medium, and may vary with time as well. The strain tensor $\boldsymbol{\varepsilon}$ merely specifies the displacement of the medium particles in the neighborhood of the point, while the stress tensor $\boldsymbol{\sigma}$ specifies the forces that neighboring parcels of the medium are exerting on each other. Therefore, they are independent of the composition and physical state of the material. The stiffness tensor $\mathbf{c}$, on the other hand, is a property of the material, and often depends on physical state variables such as temperature, pressure, and microstructure.

Due to the inherent symmetries of $\boldsymbol{\sigma}$, $\boldsymbol{\varepsilon}$, and $\mathbf{c}$, only 21 elastic coefficients of the latter are independent. For isotropic media (which have the same physical properties in any direction), $\mathbf{c}$ can be reduced to only two independent numbers, the bulk modulus K and the shear modulus G, that quantify the material's resistance to changes in volume and to shearing deformations, respectively.

Analogous Laws

Since Hooke's law is a simple proportionality between two quantities, its formulas and consequences are mathematically similar to those of many other physical laws, such as those describing the motion of fluids, or the polarization of a dielectric by an electric field.

In particular, the tensor equation $\sigma = c\varepsilon$ relating elastic stresses to strains is entirely similar to the equation $\tau = \mu\dot{\varepsilon}$ relating the viscous stress tensor τ and the strain rate tensor $\dot{\varepsilon}$ in flows of viscous fluids; although the former pertains to static stresses (related to *amount* of deformation) while the latter pertains to dynamical stresses (related to the *rate* of deformation).

Units of Measurement

In SI units, displacements are measured in meters (m), and forces in newtons (N or kg·m/s²). Therefore, the spring constant k, and each element of the tensor κ, is measured in newtons per meter (N/m), or kilograms per second squared (kg/s²).

For continuous media, each element of the stress tensor **σ** is a force divided by an area; it is therefore measured in units of pressure, namely pascals (Pa, or N/m², or kg/(m·s²)). The elements of the strain tensor **ε** are dimensionless (displacements divided by distances). Therefore, the entries of c_{ijkl} are also expressed in units of pressure.

General Application to Elastic Materials

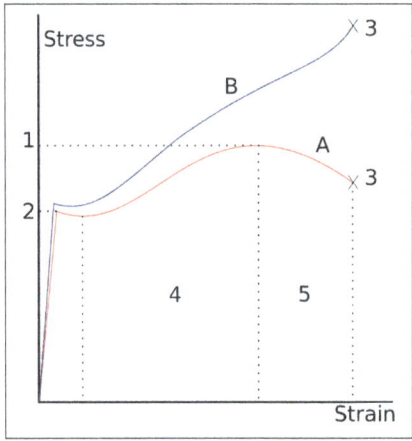

Stress–strain curve for low-carbon steel, showing the relationship between the stress (force per unit area) and strain (resulting compression/stretching, known as deformation). Hooke's law is only valid for the portion of the curve between the origin and the yield point.

- Ultimate strength.
- Yield strength (yield point).
- Rupture.
- Strain hardening region.
- Necking region.
- Apparent stress (F/A_o).
- Actual stress (F/A).

Objects that quickly regain their original shape after being deformed by a force, with the molecules or atoms of their material returning to the initial state of stable equilibrium, often obey Hooke's law.

Hooke's law only holds for some materials under certain loading conditions. Steel exhibits linear-elastic behavior in most engineering applications; Hooke's law is valid for it throughout its elastic range (i.e., for stresses below the yield strength). For some other materials, such as aluminium, Hooke's law is only valid for a portion of the elastic range. For these materials a proportional limit stress is defined, below which the errors associated with the linear approximation are negligible.

Rubber is generally regarded as a "non-Hookean" material because its elasticity is stress dependent and sensitive to temperature and loading rate.

Generalizations of Hooke's law for the case of large deformations is provided by models of neo-Hookean solids and Mooney–Rivlin solids.

Derived Formulae

Tensional Stress of a Uniform Bar

A rod of any elastic material may be viewed as a linear spring. The rod has length L and cross-sectional area A. Its tensile stress σ is linearly proportional to its fractional extension or strain ε by the modulus of elasticity E:

$$\sigma = E\varepsilon.$$

The modulus of elasticity may often be considered constant. In turn,

$$\varepsilon = \frac{\Delta L}{L}$$

(that is, the fractional change in length), and since:

$$\sigma = \frac{F}{A},$$

it follows that:

$$\varepsilon = \frac{\sigma}{E} = \frac{F}{AE}.$$

The change in length may be expressed as:

$$\Delta L = \varepsilon L = \frac{FL}{AE}.$$

Spring Energy

The potential energy $U_{el}(x)$ stored in a spring is given by:

$$U_{el}(x) = \tfrac{1}{2}kx^2$$

which comes from adding up the energy it takes to incrementally compress the spring. That is, the integral of force over displacement. Since the external force has the same general direction as the displacement, the potential energy of a spring is always non-negative.

This potential U_{el} can be visualized as a parabola on the Ux-plane such that $U_{el}(x) = 1/2kx^2$. As the spring is stretched in the positive x-direction, the potential energy increases parabolically (the same thing happens as the spring is compressed). Since the change in potential energy changes at a constant rate:

$$\frac{d^2U_{el}}{dx^2} = k.$$

Note that the change in the change in U is constant even when the displacement and acceleration are zero.

Relaxed Force Constants (Generalized Compliance Constants)

Relaxed force constants (the inverse of generalized compliance constants) are uniquely defined for molecular systems, in contradistinction to the usual "rigid" force constants, and thus their use allows meaningful correlations to be made between force fields calculated for reactants, transition states, and products of a chemical reaction. Just as the potential energy can be written as a quadratic form in the internal coordinates, so it can also be written in terms of generalized forces. The resulting coefficients are termed compliance constants. A direct method exists for calculating the compliance constant for any internal coordinate of a molecule, without the need to do the normal mode analysis. The suitability of relaxed force constants (inverse compliance constants) as covalent bond strength descriptors was demonstrated as early as 1980. Recently, the suitability as non-covalent bond strength descriptors was demonstrated too.

Harmonic Oscillator

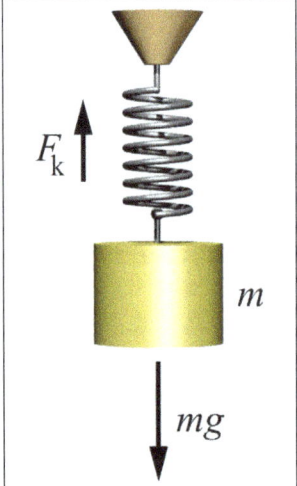

A mass suspended by a spring is the classical example of a harmonic oscillator.

A mass m attached to the end of a spring is a classic example of a harmonic oscillator. By pulling slightly on the mass and then releasing it, the system will be set in sinusoidal oscillating motion about the equilibrium position. To the extent that the spring obeys Hooke's law, and that one can neglect friction and the mass of the spring, the amplitude of the oscillation will remain constant, and its frequency f will be independent of its amplitude, determined only by the mass and the stiffness of the spring:

$$f = \frac{1}{2\pi}\sqrt{\frac{k}{m}}$$

This phenomenon made possible the construction of accurate mechanical clocks and watches that could be carried on ships and people's pockets.

Rotation in Gravity-free Space

If the mass m were attached to a spring with force constant k and rotating in free space, the spring tension (F_t) would supply the required centripetal force (F_c):

$$F_t = kx; \qquad F_c = m\omega^2 r$$

Since $F_t = F_c$ and $x = r$, then:

$$k = m\omega^2$$

Given that $\omega = 2\pi f$, this leads to the same frequency equation as above:

$$f = \frac{1}{2\pi}\sqrt{\frac{k}{m}}$$

Linear Elasticity Theory for Continuous Media

Isotropic Materials

Isotropic materials are characterized by properties which are independent of direction in space. Physical equations involving isotropic materials must therefore be independent of the coordinate system chosen to represent them. The strain tensor is a symmetric tensor. Since the trace of any tensor is independent of any coordinate system, the most complete coordinate-free decomposition of a symmetric tensor is to represent it as the sum of a constant tensor and a traceless symmetric tensor. Thus in index notation:

$$\varepsilon_{ij} = \left(\tfrac{1}{3}\varepsilon_{kk}\delta_{ij}\right) + \left(\varepsilon_{ij} - \tfrac{1}{3}\varepsilon_{kk}\delta_{ij}\right)$$

where δ_{ij} is the Kronecker delta. In direct tensor notation:

$$\varepsilon = \mathrm{vol}(\varepsilon) + \mathrm{dev}(\varepsilon); \qquad \mathrm{vol}(\varepsilon) = \tfrac{1}{3}\mathrm{tr}(\varepsilon)\mathbf{I}; \qquad \mathrm{dev}(\varepsilon) = \varepsilon - \mathrm{vol}(\varepsilon)$$

where I is the second-order identity tensor.

The first term on the right is the constant tensor, also known as the volumetric strain tensor, and the second term is the traceless symmetric tensor, also known as the deviatoric strain tensor or shear tensor.

The most general form of Hooke's law for isotropic materials may now be written as a linear combination of these two tensors:

$$\sigma_{ij} = 3K\left(\tfrac{1}{3}\varepsilon_{kk}\delta_{ij}\right) + 2G\left(\varepsilon_{ij} - \tfrac{1}{3}\varepsilon_{kk}\delta_{ij}\right); \qquad \sigma = 3K\,\mathrm{vol}(\varepsilon) + 2G\,\mathrm{dev}(\varepsilon)$$

where K is the bulk modulus and G is the shear modulus.

Using the relationships between the elastic moduli, these equations may also be expressed in various other ways. A common form of Hooke's law for isotropic materials, expressed in direct tensor notation, is:

$$\sigma = \lambda\,\mathrm{tr}(\varepsilon)\mathbf{I} + 2\mu\varepsilon = \mathbf{c}:\varepsilon; \qquad \mathbf{c} = \lambda\mathbf{I}\otimes\mathbf{I} + 2\mu\mathbf{I}$$

where $\lambda = K - 2/3G = c_{1111} - 2c_{1212}$ and $\mu = G = c_{1212}$ are the Lamé constants, I is the second-rank identity tensor, and I is the symmetric part of the fourth-rank identity tensor. In index notation:

$$\sigma_{ij} = \lambda\varepsilon_{kk}\,\delta_{ij} + 2\mu\varepsilon_{ij} = c_{ijkl}\varepsilon_{kl}; \qquad c_{ijkl} = \lambda\delta_{ij}\delta_{kl} + \mu\left(\delta_{ik}\delta_{jl} + \delta_{il}\delta_{jk}\right)$$

The inverse relationship is:

$$\varepsilon = \frac{1}{2\mu}\sigma - \frac{\lambda}{2\mu(3\lambda + 2\mu)}\mathrm{tr}(\sigma)\mathbf{I} = \frac{1}{2G}\sigma + \left(\frac{1}{9K} - \frac{1}{6G}\right)\mathrm{tr}(\sigma)\mathbf{I}$$

Therefore, the compliance tensor in the relation $\varepsilon = \mathbf{s}:\sigma$ is:

$$\mathbf{s} = -\frac{\lambda}{2\mu(3\lambda + 2\mu)}\mathbf{I}\otimes\mathbf{I} + \frac{1}{2\mu}\mathbf{I} = \left(\frac{1}{9K} - \frac{1}{6G}\right)\mathbf{I}\otimes\mathbf{I} + \frac{1}{2G}\mathbf{I}$$

In terms of Young's modulus and Poisson's ratio, Hooke's law for isotropic materials can then be expressed as:

$$\varepsilon_{ij} = \frac{1}{E}\left(\sigma_{ij} - v(\sigma_{kk}\delta_{ij} - \sigma_{ij})\right); \qquad \varepsilon = \frac{1}{E}\left(\sigma - v(\mathrm{tr}(\sigma)\mathbf{I} - \sigma)\right) = \frac{1+v}{E}\sigma - \frac{v}{E}\mathrm{tr}(\sigma)\mathbf{I}$$

This is the form in which the strain is expressed in terms of the stress tensor in engineering. The expression in expanded form is:

$$\varepsilon_{11} = \frac{1}{E}\left(\sigma_{11} - v(\sigma_{22} + \sigma_{33})\right)$$

$$\varepsilon_{22} = \frac{1}{E}\left(\sigma_{22} - v(\sigma_{11} + \sigma_{33})\right)$$

$$\varepsilon_{33} = \frac{1}{E}\left(\sigma_{33} - v(\sigma_{11} + \sigma_{22})\right)$$

$$\varepsilon_{12} = \frac{1}{2G}\sigma_{12}; \qquad \varepsilon_{13} = \frac{1}{2G}\sigma_{13}; \qquad \varepsilon_{23} = \frac{1}{2G}\sigma_{23}$$

where E is Young's modulus and v is Poisson's ratio.

In matrix form, Hooke's law for isotropic materials can be written as:

$$
\begin{bmatrix} \varepsilon_{11} \\ \varepsilon_{22} \\ \varepsilon_{33} \\ 2\varepsilon_{23} \\ 2\varepsilon_{13} \\ 2\varepsilon_{12} \end{bmatrix} = \begin{bmatrix} \varepsilon_{11} \\ \varepsilon_{22} \\ \varepsilon_{33} \\ \gamma_{23} \\ \gamma_{13} \\ \gamma_{12} \end{bmatrix} = \frac{1}{E} \begin{bmatrix} 1 & -v & -v & 0 & 0 & 0 \\ -v & 1 & -v & 0 & 0 & 0 \\ -v & -v & 1 & 0 & 0 & 0 \\ 0 & 0 & 0 & 2+2v & 0 & 0 \\ 0 & 0 & 0 & 0 & 2+2v & 0 \\ 0 & 0 & 0 & 0 & 0 & 2+2v \end{bmatrix} \begin{bmatrix} \sigma_{11} \\ \sigma_{22} \\ \sigma_{33} \\ \sigma_{23} \\ \sigma_{13} \\ \sigma_{12} \end{bmatrix}
$$

where $\gamma_{ij} = 2\varepsilon_{ij}$ is the engineering shear strain. The inverse relation may be written as:

$$
\begin{bmatrix} \sigma_{11} \\ \sigma_{22} \\ \sigma_{33} \\ \sigma_{23} \\ \sigma_{13} \\ \sigma_{12} \end{bmatrix} = \frac{E}{(1+v)(1-2v)} \begin{bmatrix} 1-v & v & v & 0 & 0 & 0 \\ v & 1-v & v & 0 & 0 & 0 \\ v & v & 1-v & 0 & 0 & 0 \\ 0 & 0 & 0 & \dfrac{1-2v}{2} & 0 & 0 \\ 0 & 0 & 0 & 0 & \dfrac{1-2v}{2} & 0 \\ 0 & 0 & 0 & 0 & 0 & \dfrac{1-2v}{2} \end{bmatrix} \begin{bmatrix} \varepsilon_{11} \\ \varepsilon_{22} \\ \varepsilon_{33} \\ 2\varepsilon_{23} \\ 2\varepsilon_{13} \\ 2\varepsilon_{12} \end{bmatrix}
$$

which can be simplified thanks to the Lamé constants:

$$
\begin{bmatrix} \sigma_{11} \\ \sigma_{22} \\ \sigma_{33} \\ \sigma_{23} \\ \sigma_{13} \\ \sigma_{12} \end{bmatrix} = \begin{bmatrix} 2\mu+\lambda & \lambda & \lambda & 0 & 0 & 0 \\ \lambda & 2\mu+\lambda & \lambda & 0 & 0 & 0 \\ \lambda & \lambda & 2\mu+\lambda & 0 & 0 & 0 \\ 0 & 0 & 0 & \mu & 0 & 0 \\ 0 & 0 & 0 & 0 & \mu & 0 \\ 0 & 0 & 0 & 0 & 0 & \mu \end{bmatrix} \begin{bmatrix} \varepsilon_{11} \\ \varepsilon_{22} \\ \varepsilon_{33} \\ 2\varepsilon_{23} \\ 2\varepsilon_{13} \\ 2\varepsilon_{12} \end{bmatrix}
$$

In vector notation this becomes:

$$
\begin{bmatrix} \sigma_{11} & \sigma_{12} & \sigma_{13} \\ \sigma_{12} & \sigma_{22} & \sigma_{23} \\ \sigma_{13} & \sigma_{23} & \sigma_{33} \end{bmatrix} = 2\mu \begin{bmatrix} \varepsilon_{11} & \varepsilon_{12} & \varepsilon_{13} \\ \varepsilon_{12} & \varepsilon_{22} & \varepsilon_{23} \\ \varepsilon_{13} & \varepsilon_{23} & \varepsilon_{33} \end{bmatrix} + \lambda \mathbf{I}(\varepsilon_{11} + \varepsilon_{22} + \varepsilon_{33})
$$

where I is the identity tensor.

Plane Stress

Under plane stress conditions, $\sigma_{31} = \sigma_{13} = \sigma_{32} = \sigma_{23} = \sigma_{33} = 0$. In that case Hooke's law takes the form:

$$
\begin{bmatrix} \varepsilon_{11} \\ \varepsilon_{22} \\ 2\varepsilon_{12} \end{bmatrix} = \frac{1}{E} \begin{bmatrix} 1 & -v & 0 \\ -v & 1 & 0 \\ 0 & 0 & 2+2v \end{bmatrix} \begin{bmatrix} \sigma_{11} \\ \sigma_{22} \\ \sigma_{12} \end{bmatrix}
$$

The inverse relation is usually written in the reduced form:

$$\begin{bmatrix} \sigma_{11} \\ \sigma_{22} \\ \sigma_{12} \end{bmatrix} = \frac{E}{1-v^2} \begin{bmatrix} 1 & v & 0 \\ v & 1 & 0 \\ 0 & 0 & \dfrac{1-v}{2} \end{bmatrix} \begin{bmatrix} \varepsilon_{11} \\ \varepsilon_{22} \\ 2\varepsilon_{12} \end{bmatrix}$$

In vector notation this becomes:

$$\begin{bmatrix} \sigma_{11} & \sigma_{12} \\ \sigma_{12} & \sigma_{22} \end{bmatrix} = \frac{E}{1-v^2} \left((1-v) \begin{bmatrix} \varepsilon_{11} & \varepsilon_{12} \\ \varepsilon_{12} & \varepsilon_{22} \end{bmatrix} + v\mathbf{I}(\varepsilon_{11} + \varepsilon_{22}) \right).$$

Anisotropic Materials

The symmetry of the Cauchy stress tensor ($\sigma_{ij} = \sigma_{ji}$ and the generalized Hooke's laws ($\sigma_{ij} = c_{ijkl}\varepsilon_{kl}$) implies that $c_{ijkl} = c_{jikl}$. Similarly, the symmetry of the infinitesimal strain tensor implies that $c_{ijkl} = c_{ijlk}$. These symmetries are called the minor symmetries of the stiffness tensor **c**. This reduces the number of elastic constants from 81 to 36.

If in addition, since the displacement gradient and the Cauchy stress are work conjugate, the stress–strain relation can be derived from a strain energy density functional (U), then:

$$\sigma_{ij} = \frac{\partial U}{\partial \varepsilon_{ij}} \quad \Rightarrow \quad c_{ijkl} = \frac{\partial^2 U}{\partial \varepsilon_{ij} \partial \varepsilon_{kl}}.$$

The arbitrariness of the order of differentiation implies that $c_{ijkl} = c_{klij}$. These are called the major symmetries of the stiffness tensor. This reduces the number of elastic constants from 36 to 21. The major and minor symmetries indicate that the stiffness tensor has only 21 independent components.

Matrix Representation (Stiffness Tensor)

It is often useful to express the anisotropic form of Hooke's law in matrix notation, also called Voigt notation. To do this we take advantage of the symmetry of the stress and strain tensors and express them as six-dimensional vectors in an orthonormal coordinate system (e_1, e_2, e_3) as:

$$[\sigma] = \begin{bmatrix} \sigma_{11} \\ \sigma_{22} \\ \sigma_{33} \\ \sigma_{23} \\ \sigma_{13} \\ \sigma_{12} \end{bmatrix} \equiv \begin{bmatrix} \sigma_1 \\ \sigma_2 \\ \sigma_3 \\ \sigma_4 \\ \sigma_5 \\ \sigma_6 \end{bmatrix} ; \quad [\varepsilon] = \begin{bmatrix} \varepsilon_{11} \\ \varepsilon_{22} \\ \varepsilon_{33} \\ 2\varepsilon_{23} \\ 2\varepsilon_{13} \\ 2\varepsilon_{12} \end{bmatrix} \equiv \begin{bmatrix} \varepsilon_1 \\ \varepsilon_2 \\ \varepsilon_3 \\ \varepsilon_4 \\ \varepsilon_5 \\ \varepsilon_6 \end{bmatrix}$$

Then the stiffness tensor (c) can be expressed as:

$$[c] = \begin{bmatrix} c_{1111} & c_{1122} & c_{1133} & c_{1123} & c_{1131} & c_{1112} \\ c_{2211} & c_{2222} & c_{2233} & c_{2223} & c_{2231} & c_{2212} \\ c_{3311} & c_{3322} & c_{3333} & c_{3323} & c_{3331} & c_{3312} \\ c_{2311} & c_{2322} & c_{2333} & c_{2323} & c_{2331} & c_{2312} \\ c_{3111} & c_{3122} & c_{3133} & c_{3123} & c_{3131} & c_{3112} \\ c_{1211} & c_{1222} & c_{1233} & c_{1223} & c_{1231} & c_{1212} \end{bmatrix} \equiv \begin{bmatrix} C_{11} & C_{12} & C_{13} & C_{14} & C_{15} & C_{16} \\ C_{12} & C_{22} & C_{23} & C_{24} & C_{25} & C_{26} \\ C_{13} & C_{23} & C_{33} & C_{34} & C_{35} & C_{36} \\ C_{14} & C_{24} & C_{34} & C_{44} & C_{45} & C_{46} \\ C_{15} & C_{25} & C_{35} & C_{45} & C_{55} & C_{56} \\ C_{16} & C_{26} & C_{36} & C_{46} & C_{56} & C_{66} \end{bmatrix}$$

and Hooke's law is written as:

$$[\sigma] = [C][\varepsilon] \quad \text{or} \quad \sigma_i = C_{ij}\varepsilon_j.$$

Similarly the compliance tensor (s) can be written as:

$$[s] = \begin{bmatrix} s_{1111} & s_{1122} & s_{1133} & 2s_{1123} & 2s_{1131} & 2s_{1112} \\ s_{2211} & s_{2222} & s_{2233} & 2s_{2223} & 2s_{2231} & 2s_{2212} \\ s_{3311} & s_{3322} & s_{3333} & 2s_{3323} & 2s_{3331} & 2s_{3312} \\ 2s_{2311} & 2s_{2322} & 2s_{2333} & 4s_{2323} & 4s_{2331} & 4s_{2312} \\ 2s_{3111} & 2s_{3122} & 2s_{3133} & 4s_{3123} & 4s_{3131} & 4s_{3112} \\ 2s_{1211} & 2s_{1222} & 2s_{1233} & 4s_{1223} & 4s_{1231} & 4s_{1212} \end{bmatrix} \equiv \begin{bmatrix} S_{11} & S_{12} & S_{13} & S_{14} & S_{15} & S_{16} \\ S_{12} & S_{22} & S_{23} & S_{24} & S_{25} & S_{26} \\ S_{13} & S_{23} & S_{33} & S_{34} & S_{35} & S_{36} \\ S_{14} & S_{24} & S_{34} & S_{44} & S_{45} & S_{46} \\ S_{15} & S_{25} & S_{35} & S_{45} & S_{55} & S_{56} \\ S_{16} & S_{26} & S_{36} & S_{46} & S_{56} & S_{66} \end{bmatrix}$$

Change of Coordinate System

If a linear elastic material is rotated from a reference configuration to another, then the material is symmetric with respect to the rotation if the components of the stiffness tensor in the rotated configuration are related to the components in the reference configuration by the relation,

$$c_{pqrs} = l_{pi}l_{qj}l_{rk}l_{sl}c_{ijkl}$$

where l_{ab} are the components of an orthogonal rotation matrix $[L]$. The same relation also holds for inversions.

In matrix notation, if the transformed basis (rotated or inverted) is related to the reference basis by:

$$[\mathbf{e}_{i'}] = [L][\mathbf{e}_i]$$

then:

$$C_{ij}\varepsilon_i\varepsilon_j = C_{i'j'}\varepsilon_i'\varepsilon_j'.$$

In addition, if the material is symmetric with respect to the transformation $[L]$ then:

$$C_{ij} = C_{ij}' \quad \Rightarrow \quad C_{ij}(\varepsilon_i\varepsilon_j - \varepsilon_i'\varepsilon_j') = 0.$$

Orthotropic Materials

Orthotropic materials have three orthogonal planes of symmetry. If the basis vectors (e_1, e_2, e_3) are normals to the planes of symmetry then the coordinate transformation relations imply that:

$$
\begin{bmatrix} \sigma_1 \\ \sigma_2 \\ \sigma_3 \\ \sigma_4 \\ \sigma_5 \\ \sigma_6 \end{bmatrix} =
\begin{bmatrix}
C_{11} & C_{12} & C_{13} & 0 & 0 & 0 \\
C_{12} & C_{22} & C_{23} & 0 & 0 & 0 \\
C_{13} & C_{23} & C_{33} & 0 & 0 & 0 \\
0 & 0 & 0 & C_{44} & 0 & 0 \\
0 & 0 & 0 & 0 & C_{55} & 0 \\
0 & 0 & 0 & 0 & 0 & C_{66}
\end{bmatrix}
\begin{bmatrix} \varepsilon_1 \\ \varepsilon_2 \\ \varepsilon_3 \\ \varepsilon_4 \\ \varepsilon_5 \\ \varepsilon_6 \end{bmatrix}
$$

The inverse of this relation is commonly written as:

$$
\begin{bmatrix} \varepsilon_{xx} \\ \varepsilon_{yy} \\ \varepsilon_{zz} \\ 2\varepsilon_{yz} \\ 2\varepsilon_{zx} \\ 2\varepsilon_{xy} \end{bmatrix} =
\begin{bmatrix}
\dfrac{1}{E_x} & -\dfrac{v_{yx}}{E_y} & -\dfrac{v_{zx}}{E_z} & 0 & 0 & 0 \\[2mm]
-\dfrac{v_{xy}}{E_x} & \dfrac{1}{E_y} & -\dfrac{v_{zy}}{E_z} & 0 & 0 & 0 \\[2mm]
-\dfrac{v_{xz}}{E_x} & -\dfrac{v_{yz}}{E_y} & \dfrac{1}{E_z} & 0 & 0 & 0 \\[2mm]
0 & 0 & 0 & \dfrac{1}{G_{yz}} & 0 & 0 \\[2mm]
0 & 0 & 0 & 0 & \dfrac{1}{G_{zx}} & 0 \\[2mm]
0 & 0 & 0 & 0 & 0 & \dfrac{1}{G_{xy}}
\end{bmatrix}
\begin{bmatrix} \sigma_{xx} \\ \sigma_{yy} \\ \sigma_{zz} \\ \sigma_{yz} \\ \sigma_{zx} \\ \sigma_{xy} \end{bmatrix}
$$

where,

E_i is the Young's modulus along axis i.

G_{ij} is the shear modulus in direction j on the plane whose normal is in direction i.

v_{ij} is the Poisson's ratio that corresponds to a contraction in direction j when an extension is applied in direction i.

Under *plane stress* conditions, $\sigma_{zz} = \sigma_{zx} = \sigma_{yz} = 0$, Hooke's law for an orthotropic material takes the form:

$$
\begin{bmatrix} \varepsilon_{xx} \\ \varepsilon_{yy} \\ 2\varepsilon_{xy} \end{bmatrix} =
\begin{bmatrix}
\dfrac{1}{E_x} & -\dfrac{v_{yx}}{E_y} & 0 \\[2mm]
-\dfrac{v_{xy}}{E_x} & \dfrac{1}{E_y} & 0 \\[2mm]
0 & 0 & \dfrac{1}{G_{xy}}
\end{bmatrix}
\begin{bmatrix} \sigma_{xx} \\ \sigma_{yy} \\ \sigma_{xy} \end{bmatrix}.
$$

The inverse relation is:

$$
\begin{bmatrix} \sigma_{xx} \\ \sigma_{yy} \\ \sigma_{xy} \end{bmatrix} = \frac{1}{1-v_{xy}v_{yx}} \begin{bmatrix} E_x & v_{yx}E_x & 0 \\ v_{xy}E_y & E_y & 0 \\ 0 & 0 & G_{xy}(1-v_{xy}v_{yx}) \end{bmatrix} \begin{bmatrix} \varepsilon_{xx} \\ \varepsilon_{yy} \\ 2\varepsilon_{xy} \end{bmatrix}.
$$

The transposed form of the above stiffness matrix is also often used.

Transversely Isotropic Materials

A transversely isotropic material is symmetric with respect to a rotation about an axis of symmetry. For such a material, if e_3 is the axis of symmetry, Hooke's law can be expressed as:

$$
\begin{bmatrix} \sigma_1 \\ \sigma_2 \\ \sigma_3 \\ \sigma_4 \\ \sigma_5 \\ \sigma_6 \end{bmatrix} = \begin{bmatrix} C_{11} & C_{12} & C_{13} & 0 & 0 & 0 \\ C_{12} & C_{11} & C_{13} & 0 & 0 & 0 \\ C_{13} & C_{13} & C_{33} & 0 & 0 & 0 \\ 0 & 0 & 0 & C_{44} & 0 & 0 \\ 0 & 0 & 0 & 0 & C_{44} & 0 \\ 0 & 0 & 0 & 0 & 0 & \dfrac{C_{11}-C_{12}}{2} \end{bmatrix} \begin{bmatrix} \varepsilon_1 \\ \varepsilon_2 \\ \varepsilon_3 \\ \varepsilon_4 \\ \varepsilon_5 \\ \varepsilon_6 \end{bmatrix}
$$

More frequently, the $x \equiv e_1$ axis is taken to be the axis of symmetry and the inverse Hooke's law is written as:

$$
\begin{bmatrix} \varepsilon_{xx} \\ \varepsilon_{yy} \\ \varepsilon_{zz} \\ 2\varepsilon_{yz} \\ 2\varepsilon_{zx} \\ 2\varepsilon_{xy} \end{bmatrix} = \begin{bmatrix} \dfrac{1}{E_x} & -\dfrac{v_{yx}}{E_y} & -\dfrac{v_{yx}}{E_y} & 0 & 0 & 0 \\ -\dfrac{v_{xy}}{E_x} & \dfrac{1}{E_y} & -\dfrac{v_{yz}}{E_y} & 0 & 0 & 0 \\ -\dfrac{v_{xy}}{E_x} & -\dfrac{v_{yz}}{E_y} & \dfrac{1}{E_y} & 0 & 0 & 0 \\ 0 & 0 & 0 & \dfrac{2+2v_{yz}}{E_y} & 0 & 0 \\ 0 & 0 & 0 & 0 & \dfrac{1}{G_{xy}} & 0 \\ 0 & 0 & 0 & 0 & 0 & \dfrac{1}{G_{xy}} \end{bmatrix} \begin{bmatrix} \sigma_{xx} \\ \sigma_{yy} \\ \sigma_{zz} \\ \sigma_{yz} \\ \sigma_{zx} \\ \sigma_{xy} \end{bmatrix}
$$

Universal Elastic Anisotropy Index

To grasp the degree of anisotropy of any class, a Universal Elastic Anisotropy Index (AU) was formulated. It replaces the Zener ratio, which is suited for cubic crystals.

Thermodynamic Basis

Linear deformations of elastic materials can be approximated as adiabatic. Under these conditions and for quasistatic processes the first law of thermodynamics for a deformed body can be expressed as:

$$\delta W = \delta U$$

where δU is the increase in internal energy and δW is the work done by external forces. The work can be split into two terms:

$$\delta W = \delta W_s + \delta W_b$$

where δW_s is the work done by surface forces while δW_b is the work done by body forces. If $\delta \mathbf{u}$ is a variation of the displacement field $\mathbf{u}$ in the body, then the two external work terms can be expressed as:

$$\delta W_s = \int_{\partial \Omega} \mathbf{t} \cdot \delta \mathbf{u} \, dS; \qquad \delta W_b = \int_{\Omega} \mathbf{b} \cdot \delta \mathbf{u} \, dV$$

where t is the surface traction vector, b is the body force vector, Ω represents the body and $\partial\Omega$ represents its surface. Using the relation between the Cauchy stress and the surface traction, t = n · σ (where n is the unit outward normal to $\partial\Omega$), we have:

$$\delta W = \delta U = \int_{\partial\Omega} (\mathbf{n} \cdot \sigma) \cdot \delta \mathbf{u} \, dS + \int_{\Omega} \mathbf{b} \cdot \delta \mathbf{u} \, dV.$$

Converting the surface integral into a volume integral via the divergence theorem gives:

$$\delta U = \int_{\Omega} \left(\nabla \cdot (\sigma \cdot \delta \mathbf{u}) + \mathbf{b} \cdot \delta \mathbf{u} \right) dV.$$

Using the symmetry of the Cauchy stress and the identity:

$$\nabla \cdot (\mathbf{a} \cdot \mathbf{b}) = (\nabla \cdot \mathbf{a}) \cdot \mathbf{b} + \tfrac{1}{2} \left(\mathbf{a}^\mathsf{T} : \nabla \mathbf{b} + \mathbf{a} : (\nabla \mathbf{b})^\mathsf{T} \right)$$

we have the following:

$$\delta U = \int_{\Omega} \left(\sigma : \tfrac{1}{2} \left(\nabla \delta \mathbf{u} + (\nabla \delta \mathbf{u})^\mathsf{T} \right) + (\nabla \cdot \sigma + \mathbf{b}) \cdot \delta \mathbf{u} \right) dV.$$

From the definition of strain and from the equations of equilibrium we have:

$$\delta \varepsilon = \tfrac{1}{2} \left(\nabla \delta \mathbf{u} + (\nabla \delta \mathbf{u})^\mathsf{T} \right); \qquad \nabla \cdot \sigma + \mathbf{b} = 0.$$

Hence we can write:

$$\delta U = \int_{\Omega} \sigma : \delta \varepsilon \, dV$$

and therefore the variation in the internal energy density is given by:

$$\delta U_0 = \sigma : \delta \varepsilon.$$

An elastic material is defined as one in which the total internal energy is equal to the potential energy of the internal forces (also called the elastic strain energy). Therefore, the internal energy density is a function of the strains, $U_0 = U_0(\varepsilon)$ and the variation of the internal energy can be expressed as:

$$\delta U_0 = \frac{\partial U_0}{\partial \varepsilon} : \delta \varepsilon.$$

Since the variation of strain is arbitrary, the stress–strain relation of an elastic material is given by:

$$\sigma = \frac{\partial U_0}{\partial \varepsilon}.$$

For a linear elastic material, the quantity $\partial U_{0/\partial}\varepsilon$ is a linear function of ε, and can therefore be expressed as:

$$\sigma = C : \varepsilon$$

where **c** is a fourth-rank tensor of material constants, also called the stiffness tensor. We can see why **c** must be a fourth-rank tensor by noting that, for a linear elastic material:

$$\frac{\partial}{\partial \varepsilon} \sigma(\varepsilon) = \text{constant} = C.$$

In index notation:

$$\frac{\partial \sigma_{ij}}{\partial \varepsilon_{kl}} = \text{constant} = c_{ijkl}.$$

The right-hand side constant requires four indices and is a fourth-rank quantity. We can also see that this quantity must be a tensor because it is a linear transformation that takes the strain tensor to the stress tensor. We can also show that the constant obeys the tensor transformation rules for fourth-rank tensors.

BOYLE'S LAW

Boyle's law, most often referred to as the Boyle–Mariotte law, or Mariotte's law (especially in France), is an experimental gas law that describes how the pressure of a gas tends to increase as the volume of the container decreases. A modern statement of Boyle's law is

The absolute pressure exerted by a given mass of an ideal gas is inversely proportional to the volume it occupies if the temperature and amount of gas remain unchanged within a closed system.

Mathematically, Boyle's law can be stated as:

$P \propto \dfrac{1}{V}$ Pressure is inversely proportional to the volume.

or

$PV = k$ Pressure multiplied by volume equals some constant k.

where P is the pressure of the gas, V is the volume of the gas, and k is a constant.

The equation states that the product of pressure and volume is a constant for a given mass of confined gas and this holds as long as the temperature is constant. For comparing the same substance under two different sets of conditions, the law can be usefully expressed as

$P_1V_1 = P_2V_2$.

This equation shows that, as volume increases, the pressure of the gas decreases in proportion. Similarly, as volume decreases, the pressure of the gas increases. The law was named after chemist and physicist Robert Boyle, who published the original law in 1662.

The law itself can be stated as follows:

For a fixed amount of an ideal gas kept at a fixed temperature, pressure and volume are inversely proportional.

Or Boyle's law is a gas law, stating that the pressure and volume of a gas have an inverse relationship, when temperature is held constant. If volume increases, then pressure decreases and vice versa, when temperature is held constant.

Therefore, when the volume is halved, the pressure is doubled; and if the volume is doubled, the pressure is halved.

Relation with Kinetic Theory and Ideal Gases

Boyle's law states that *at constant temperature* the volume of a given mass of a dry gas is inversely proportional to its pressure.

Most gases behave like ideal gases at moderate pressures and temperatures. The technology of the 17th century could not produce very high pressures or very low temperatures. Hence, the law was not likely to have deviations at the time of publication. As improvements in technology permitted higher pressures and lower temperatures, deviations from the ideal gas behavior became noticeable, and the relationship between pressure and volume can only be accurately described employing real gas theory. The deviation is expressed as the compressibility factor.

Boyle (and Mariotte) derived the law solely by experiment. The law can also be derived theoretically based on the presumed existence of atoms and molecules and assumptions about motion and perfectly elastic collisions. These assumptions were met with enormous resistance in the positivist scientific community at the time however, as they were seen as purely theoretical constructs for which there was not the slightest observational evidence.

Daniel Bernoulli in 1737-1738 derived Boyle's law by applying Newton's laws of motion at the molecular level. It remained ignored until around 1845, when John Waterston published a paper building the main precepts of kinetic theory; this was rejected by the Royal Society of England. Later works of James Prescott Joule, Rudolf Clausius and in particular Ludwig Boltzmann firmly established the kinetic theory of gases and brought attention to both the theories of Bernoulli and Waterston.

The debate between proponents of energetics and atomism led Boltzmann to write a book in 1898, which endured criticism until his suicide in 1906. Albert Einstein in 1905 showed how kinetic theory applies to the Brownian motion of a fluid-suspended particle, which was confirmed in 1908 by Jean Perrin.

Equation

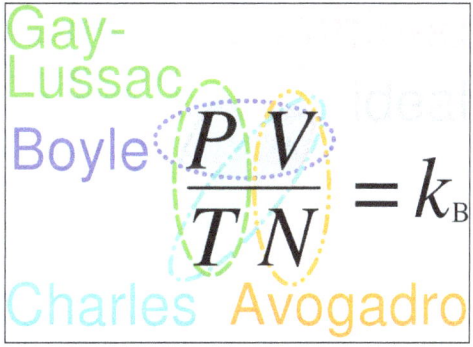

Relationships between Boyle's, Charles's, Gay-Lussac's, Avogadro's, combined and ideal gas laws, with the Boltzmann constant $k_B = R/N_A = nR/N$ (in each law, properties circled are variable and properties not circled are constant).

The mathematical equation for Boyle's law is:

$$PV = k$$

where:

- P denotes the pressure of the system.

- V denotes the volume of the gas.

- k is a constant value representative of the temperature and volume of the system.

So long as temperature remains constant the same amount of energy given to the system persists throughout its operation and therefore, theoretically, the value of k will remain constant. However, due to the derivation of pressure as perpendicular applied force and the probabilistic likelihood of collisions with other particles through collision theory, the application of force to a surface may not be infinitely constant for such values of v, but will have a limit when differentiating such values over a given time. Forcing the volume V of the fixed quantity of gas to increase, keeping the gas at the initially measured temperature, the pressure p must decrease proportionally. Conversely, reducing the volume of the gas increases the pressure. Boyle's law is used to predict the result of introducing a change, in volume and pressure only, to the initial state of a fixed quantity of gas.

The initial and final volumes and pressures of the fixed amount of gas, where the initial and final temperatures are the same (heating or cooling will be required to meet this condition), are related by the equation:

$$P_1V_1 = P_2V_2.$$

Here P_1 and V_1 represent the original pressure and volume, respectively, and P_2 and V_2 represent the second pressure and volume.

Boyle's law, Charles's law, and Gay-Lussac's law form the combined gas law. The three gas laws in combination with Avogadro's law can be generalized by the ideal gas law.

Human Breathing System

Boyle's law is often used as part of an explanation on how the breathing system works in the human body. This commonly involves explaining how the lung volume may be increased or decreased and thereby cause a relatively lower or higher air pressure within them (in keeping with Boyle's law). This forms a pressure difference between the air inside the lungs and the environmental air pressure, which in turn precipitates either inhalation or exhalation as air moves from high to low pressure.

CHARLES'S LAW

Charles' law (also known as the law of volumes) is an experimental gas law that describes how gases tend to expand when heated. A modern statement of Charles's law is:

When the pressure on a sample of a dry gas is held constant, the Kelvin temperature and the volume will be in direct proportion.

This relationship of direct proportion can be written as:

$$V \propto T$$

So this means:

$$\frac{V}{T} = k, \quad or \quad V = kT$$

where,

      V is the volume of the gas,

      T is the temperature of the gas (measured in kelvins),

      and k is a non-zero constant.

This law describes how a gas expands as the temperature increases; conversely, a decrease in temperature will lead to a decrease in volume. For comparing the same substance under two different sets of conditions, the law can be written as:

$$\frac{V_1}{T_1} = \frac{V_2}{T_2} \quad or \quad \frac{V_2}{V_1} = \frac{T_2}{T_1} \quad or \quad V_1T_2 = V_2T_1.$$

The equation shows that, as absolute temperature increases, the volume of the gas also increases in proportion.

The law was named after scientist Jacques Charles, who formulated the original law in his unpublished work from the 1780s.

In two of a series of four essays presented between 2 and 30 October 1801, John Dalton demonstrated by experiment that all the gases and vapours that he studied expanded by the same amount between two fixed points of temperature. The French natural philosopher Joseph Louis Gay-Lussac confirmed the discovery in a presentation to the French National Institute on 31 Jan 1802, although he credited the discovery to unpublished work from the 1780s by Jacques Charles. The basic principles had already been described by Guillaume Amontons and Francis Hauksbee a century earlier.

Dalton was the first to demonstrate that the law applied generally to all gases, and to the vapours of volatile liquids if the temperature was well above the boiling point. Gay-Lussac concurred. With measurements only at the two thermometric fixed points of water, Gay-Lussac was unable to show that the equation relating volume to temperature was a linear function. On mathematical grounds alone, Gay-Lussac's paper does not permit the assignment of any law stating the linear relation. Both Dalton's and Gay-Lussac's main conclusions can be expressed mathematically as:

$$V_{100} - V_0 = kV_0$$

where V_{100} is the volume occupied by a given sample of gas at 100 °C; V_0 is the volume occupied by the same sample of gas at 0 °C; and k is a constant which is the same for all gases at constant pressure. This equation does not contain the temperature and so has nothing to do with what became known as Charles' Law. Gay-Lussac's value for k ($\frac{1}{2.6666}$), was identical to Dalton's earlier value for vapours and remarkably close to the present-day value of $\frac{1}{2.7315}$. Gay-Lussac gave credit for this equation to unpublished statements by his fellow Republican citizen J. Charles in 1787. In the absence of a firm record, the gas law relating volume to temperature cannot be named after Charles. Dalton's measurements had much more scope regarding temperature than Gay-Lussac, not only measuring the volume at the fixed points of water, but also at two intermediate points. Unaware of the inaccuracies of mercury thermometers at the time, which were divided into equal portions between the fixed points, Dalton, after concluding in Essay II that in the case of vapours, "any elastic fluid expands nearly in a uniform manner into 1370 or 1380 parts by 180 degrees (Fahrenheit) of heat", was unable to confirm it for gases.

Relation to Absolute Zero

Charles' law appears to imply that the volume of a gas will descend to zero at a certain temperature (−266.66 °C according to Gay-Lussac's figures) or −273.15 °C. Gay-Lussac was clear in his description that the law was not applicable at low temperatures:

> but I may mention that this last conclusion cannot be true except so long as the compressed vapours remain entirely in the elastic state; and this requires that their temperature shall be sufficiently elevated to enable them to resist the pressure which tends to make them assume the liquid state.

At absolute zero temperature the gas possesses zero energy and hence the molecules restrict motion. Gay-Lussac had no experience of liquid air, although he appears to have believed (as did Dalton) that the "permanent gases" such as air and hydrogen could be liquified. Gay-Lussac had also worked with the vapours of volatile liquids in demonstrating Charles' law, and was aware that the law does not apply just above the boiling point of the liquid:

> I may however remark that when the temperature of the ether is only a little above its boiling point, its condensation is a little more rapid than that of atmospheric air. This fact is related to a phenomenon which is exhibited by a great many bodies when passing from the liquid to the solid state, but which is no longer sensible at temperatures a few degrees above that at which the transition occurs.

The first mention of a temperature at which the volume of a gas might descend to zero was by William Thomson (later known as Lord Kelvin) in 1848:

> This is what we might anticipate, when we reflect that infinite cold must correspond to a finite number of degrees of the air-thermometer below zero; since if we push the strict principle of graduation, stated above, sufficiently far, we should arrive at a point corresponding to the volume of air being reduced to nothing, which would be marked as −273° of the scale (−100/.366, if .366 be the coefficient of expansion); and therefore −273° of the air-thermometer is a point which cannot be reached at any finite temperature, however low.

However, the "absolute zero" on the Kelvin temperature scale was originally defined in terms of the second law of thermodynamics, which Thomson himself described in 1852. Thomson did not assume that this was equal to the "zero-volume point" of Charles' law, merely that Charles' law provided the minimum temperature which could be attained. The two can be shown to be equivalent by Ludwig Boltzmann's statistical view of entropy.

However, Charles also stated:

> The volume of a fixed mass of dry gas increases or decreases by $\frac{1}{273}$ times the volume at 0 °C for every 1 °C rise or fall in temperature.

Thus:

$$V_T = V_0 + \left(\tfrac{1}{273} \times V_0\right) \times T$$

$$V_T = V_0\left(1 + \frac{T}{273}\right)$$

where, VT is the volume of gas at temperature T, Vo is the volume at 0 °C.

Relation to Kinetic Theory

The kinetic theory of gases relates the macroscopic properties of gases, such as pressure and volume, to the microscopic properties of the molecules which make up the gas, particularly the mass and speed of the molecules. In order to derive Charles' law from kinetic theory, it is necessary to have a microscopic definition of temperature: this can be conveniently taken as the temperature being proportional to the average kinetic energy of the gas molecules, E_k.

$$T \propto \overline{E_k}.$$

Under this definition, the demonstration of Charles' law is almost trivial. The kinetic theory equivalent of the ideal gas law relates PV to the average kinetic energy:

$$PV = \frac{2}{3} N \overline{E_k}$$

References

- Milton, Graeme W. (2002). The Theory of Composites. Cambridge Monographs on Applied and Computational Mathematics. Cambridge University Press. ISBN 9780521781251

- Stefan-Boltzmann-law, Science: britannica.com, Retrieved 20 August, 2019

- Ranganathan, S.I.; Ostoja-Starzewski, M. (2008). "Universal Elastic Anisotropy Index". Physical Review Letters. 101 (5): 055504–1–4. Bibcode:2008PhRvL.101e5504R. doi:10.1103/PhysRevLett.101.055504. PMID 18764407

- Physics, chapter, 11-5-pascals-principle: courses.lumenlearning.com, Retrieved 21 January, 2019

- Gerald James Holton (2001). Physics, the Human Adventure: From Copernicus to Einstein and Beyond. Rutgers University Press. Pp. 270–. ISBN 978-0-8135-2908-0

Permissions

Index